普通高等教育"十二五"规划教材

信号与系统

崔畅　赵强　佟慧艳　主编

中国石化出版社

内 容 提 要

本书主要研究线性时不变系统传输与处理确定性信号方面的基本概念和基本分析方法。全书共分为6章，主要内容包括信号与系统概述；连续时间信号与系统的时域、频域和复频域分析；离散时间信号与系统的时域和 z 域分析。

本书适合于渗透式双语教学。书中章节标题、重要的技术术语、概念等内容均采用中英文双语形式，例题、练习题几乎采用纯英文形式，使学生在学习本课程的同时，提高阅读本专业英文书籍和文献的能力。书中还引入 MATLAB 软件作为信号与系统分析的工具，加深学生对信号与系统基本原理、方法及应用的理解。

本书可作为高等学校电气信息类等相关专业本科生教材，也可作为从事相关领域工作的教师、科技工作者的参考书。

图书在版编目(CIP)数据

信号与系统/崔畅，赵强，佟慧艳主编. —北京：
中国石化出版社，2015.1(2018.2重印)
普通高等教育"十二五"规划教材
ISBN 978-7-5114-3061-8

Ⅰ. ①信… Ⅱ. ①崔… ②赵… ③佟… Ⅲ. ①信号系统-高等学校-教材 Ⅳ. ①TN911.6

中国版本图书馆 CIP 数据核字(2014)第 277625 号

未经本社书面授权，本书任何部分不得被复制、抄袭，或者以任何形式或任何方式传播。版权所有，侵权必究。

中国石化出版社出版发行
地址：北京市朝阳区吉市口路9号
邮编：100020　电话：(010)59964500
发行部电话：(010)59964526
http://www.sinopec-press.com
E-mail:press@sinopec.com
北京艾普海德印刷有限公司印刷
全国各地新华书店经销

*

787×1092 毫米 16 开本 13.25 印张 316 千字
2015 年 1 月第 1 版　2018 年 2 月第 2 次印刷
定价:28.00元

前言(Preface)

信号与系统课程是电气信息类专业的一门重要技术基础课程，其中的概念和分析方法广泛应用于通信、自动控制、信号与信息处理、电路与系统等领域。

本书主要研究确定性信号和线性时不变系统的基本理论和方法，研究对象包括连续和离散时间信号与系统，研究方法包括时域分析和变换域分析，重点是变换域分析。本书按照先信号后系统，先连续后离散，先时域分析后变换域分析的顺序，对信号与系统的分析方法进行了全面地介绍。同时，根据教学实践中的经验与教学需要，对教材内容作了精心编排，以期能够更好地为高校信号与系统课程的教学服务。

本书内容深入浅出，通俗易懂。书中配有大量的例题和练习题，注重难点和重点的解释与分析。最重要的特点是适合于双语教学：章节标题、重要的技术术语、概念等内容均采用中英文双语形式，例题、练习题几乎采用纯英文形式，同时，在每章的最后还列出重要术语和概念的中英文对照表，使学生在学习专业知识的同时，增加本专业的英语词汇量，更好地与国际接轨，满足双语教学的需求。书中还引入MATLAB软件作为信号与系统分析的工具，将课程中的重点、难点用MATLAB进行形象、直观的可视化模拟与仿真实现，从而加深对信号与系统基本原理、方法及应用的理解。

本书共分为6章。第1章介绍了信号与系统的基本概念和基本理论；第2~4章着重讨论连续时间信号与系统的时域、频域和复频域分析；第5章和第6章分别研究了离散时间信号与系统的时域和z域分析。

本书由辽宁石油化工大学崔畅、赵强、佟慧艳主编，可作为高等学校电气信息类专业本科生教材，也可作为从事相关领域工作的教师、科技工作者的参考书。

由于时间及作者水平有限，书中难免有错误不当之处，恳请广大读者批评指正。

编者

目录(Contents)

第1章 信号与系统概述
(Introduction of Signals and Systems) ·· (1)
1.1 信息、信号和系统(Information, Signal and System) ················ (1)
1.2 信号的分类(Classification of Signals) ····································· (2)
1.3 基本的连续时间信号(Basic Continuous-time Signals) ············· (4)
1.4 基本的离散时间信号(Basic Discrete-time Signals——sequence) (12)
1.5 连续时间信号的基本变换(Basic Transformation of Continuous-time Signals) ··· (16)
1.6 系统的描述及分类(The Description and Classification of Systems) ··· (20)
1.7 典型信号的 MATLAB 实现(MATLAB Realization of Typical Signals) ··· (24)
关键词(Key Words and Phrases) ·· (26)
Exercises ··· (28)

第2章 连续时间系统的时域分析
(Analysis of Continuous-time System in Time Domain) ············· (30)
2.1 系统的时域模型(Time-domain Models of System) ················· (30)
2.2 经典时域解法(The Classical Solution in Time Domain) ········· (32)
2.3 LTI 连续时间系统的响应(The Response of LTI Continuous-time System) ····· (36)
2.4 卷积积分(The Convolution Integral) ····································· (41)
2.5 利用 MATLAB 进行连续时间系统的时域分析(Analysis of Continuous-time Systems in Time Domain Based on MATLAB) ··· (49)
关键词(Key Words and Phrases) ·· (52)
Exercises ··· (53)

第3章 连续时间信号与系统的频域分析
(Analysis of Continuous-time Signals and Systems in Frequency Domain) ··········· (56)
3.1 周期信号的频域分析(Analysis of Periodic Signals in Frequency Domain) ······ (56)
3.2 非周期信号的傅里叶变换(Fourier Transform of Aperiodic Signals) ··········· (64)
3.3 傅里叶变换的性质(Properties of Fourier Transform) ············· (71)
3.4 周期信号的傅里叶变换(Fourier Transform of Periodic Signals) ··········· (82)
3.5 连续信号的抽样定理(The Sampling Theorem for Continuous Signal) ········· (84)
3.6 LTI 系统的频域分析(Fourier Analysis of Continuous-time LTI Systems) ··· (88)
3.7 无失真传输系统(Distortionless Transmission System) ············· (92)
3.8 理想低通滤波器(Ideal Lowpass Filter) ····································· (94)
3.9 利用 MATLAB 进行连续时间信号与系统的频域分析(Fourier Analysis of Continuous-time Signals and Systems Using MATLAB) ··· (99)
关键词(Key Words and Phrases) ·· (103)
Exercises ··· (104)

I

第4章 连续时间信号与系统的复频域分析
(Analysis of Continuous-time Signals and Systems in Complex Frequency Domain) ……(108)
    4.1 拉普拉斯变换(The Laplace Transform)……(108)
    4.2 常用信号的拉普拉斯变换(Some Laplace Transform Pairs)……(112)
    4.3 拉普拉斯变换的性质(Properties of the Laplace Transform)……(114)
    4.4 拉普拉斯反变换(The Inverse Laplace Transform)……(123)
    4.5 连续 LTI 系统的复频域分析(Analysis of LTI Continuous-time System in Complex Frequency Domain)……(129)
    4.6 系统函数与系统特性(System Function and System Characteristic)……(133)
    4.7 LTI 连续时间系统的模拟(Imitation of LTI Continuous-time Systems)……(143)
    4.8 利用 MATLAB 进行连续时间系统的复频域分析(MATLAB Analysis of Continuous-time System in Complex Frequency Domain)……(146)
    关键词(Key Words and Phrases)……(149)
    Exercises ……(150)

第5章 离散时间系统的时域分析
(Analysis of the Discrete-time Systems in Time Domain) ……(153)
    5.1 离散时间系统的描述——差分方程(Description of the Discrete-time Systems——Difference Equation)……(153)
    5.2 离散时间系统的经典解法(Classical Solution of Discrete-time Systems)……(156)
    5.3 离散时间系统的响应(The Response of the Discrete-time System)……(160)
    5.4 卷积和(The Convolution Sum)……(163)
    5.5 利用 MATLAB 进行离散时间系统的时域分析(Using MATLAB to Analyse Discrete-time Systems in Time Domain)……(167)
    关键词(Key Words and Phrases)……(171)
    Exercises ……(172)

第6章 离散时间信号与系统的 z 域分析
(Analysis of Discrete-time Signals and Systems in z-Domain) ……(174)
    6.1 z 变换的定义及其收敛域(Definition and the Region of Convergence of z-transform)……(174)
    6.2 常用序列的 z 变换(The z-transform of Basic Sequence)……(177)
    6.3 z 变换的性质(Properties of the z-transform)……(179)
    6.4 z 反变换(The Inversion z-transform)……(185)
    6.5 利用 z 变换求解差分方程(Solving Difference Equation by Using z-transform)……(190)
    6.6 离散时间系统的系统函数与频率响应(System Function and Frequency Response of Discrete-time System)……(192)
    6.7 利用 MATLAB 进行离散系统的复频域分析(Analysis of Discrete-time Systems in z-Domain Based on MATLAB)……(198)
    关键词(Key Words and Phrases)……(202)
    Exercises ……(202)

第1章 信号与系统概述
(Introduction of Signals and Systems)

本章介绍信号与系统的基本概念和基本理论，是全书的基础。在信号方面概要介绍信号的描述、分类和基本变换，详细阐述了常用的典型信号、奇异信号的基本性质；在系统方面概要介绍了系统的概念和描述方法及系统的分类。

1.1 信息、信号和系统 (Information, Signal and System)

"信号(signal)"一词在人们的日常生活与社会生活中有着广泛的含义。在掌握信号的概念之前先来了解一下消息和信息的概念。

消息(message)是表达客观物质运动和主观思维活动的状态，通常用语言、文字、图像、数据等来表达。例如，电话中的声音、电视中的图像、雷达探测到的目标距离等都是消息。

信息(information)则是消息中有意义的内容。在得到一个消息后，可能得到一定数量的信息。形式上传输消息，实质上传输信息；消息具体，信息抽象；消息是表达信息的工具，信息载荷在消息中，同一信息可用不同形式的消息来载荷；消息可能包含丰富的信息，也可能包含很少的信息。

消息的传送一般都不是直接的，必须借助于一定形式的信号才能进行远距离传输和各种处理。

信号就是运载消息的工具，是消息的载体，是带有信息的某种物理量，例如，电信号、光信号、声音信号等。这些物理量包含着信息，因此，信号可以是随时间变化或空间变化的物理量，在数学上可以用一个或几个独立变量(independent variable)的函数表示，也可以用曲线、图形等方式表示。

Signals are physical phenomena or physical quantities, which change with time or space. They can represented mathematically as functions of one or more independent variables.

在可以作为信号的诸多物理量中，电是应用最广的物理量。电易于产生和控制，传输速率快，也容易实现与非电量的相互转换。因此，本书主要对电信号(electrical signal)展开讨论。电信号通常是随时间变化的电压(voltage)或电流(current)。由于是随时间而变化的，在数学上常用时间 t 的函数来表示，所以本书中"信号(signal)"与"函数(function)"这两个名词常交替使用。

In this book, we focus our attention on signals involving a single independent variable. For convenience, we will generally refer to the independent variable as time, although it may not in fact represent time in specific applications.

系统(system)是由若干相互作用和相互依赖的事物组合而成的、具有特定功能的整体。系统是由各个不同单元按照一定的方式组成并完成某种任务整体的总称。系统所完成的任务就是处理、传输和存储信号，以达到自然界、人类社会、生产设备按照对人类有利的规律运动的目的，故系统的组成、特性应由信息和信号决定。

A system is an interconnection of components (e.g. devices or processes) with terminals or access ports through which matter, energy, or information can be applied or extracted. It will often turn out to be more convenient to use the concept of the signal and the resulting response to describe the characteristics of a system. Actually, the system will sometimes only be known in terms of its response to given signals.

信号与系统是两个既相互联系又相互区别的研究对象。信号是运载信息的工具，系统是产生、传输和处理信号的客观实体。信号离开了系统就失去了存在的价值，系统没有信号的输入，也失去了作用。例如，电视信号与电视机的关系。

综上所述，信息、信号和系统是不可分割的整体。

1.2 信号的分类（Classification of Signals）

信号的分类方法有很多，可以从不同的角度对信号进行分类。例如，按实际用途可将信号划分为电视信号、雷达信号、控制信号、通信信号、广播信号等；在信号与系统分析中，按照信号所具有的时间特性可划分为确定性信号与随机信号、连续时间信号与离散时间信号、周期信号与非周期信号、能量信号与功率信号等。

1.2.1 确定性信号与随机信号（Deterministic and Random Signals）

确定性信号（deterministic signal）是对于指定的某一时刻 t，可确定相应的函数值 $f(t)$ 与之对应（有限个不连续点除外）。随机信号（random signal）具有未可预知的不确定性，不能以明确的数学表达式表示，只能知道该信号的统计特性。

A deterministic signal can be represented by distinct mathematical expressions, but a random signal cannot find a function to represent it.

两者的区分要点是：给定的自变量是否对应唯一且确定的信号取值。

本书只涉及确定性信号。

1.2.2 连续时间信号与离散时间信号（Continuous-time and Discrete-time Signals）

按照信号在时间轴上取值是否连续，可将信号分为连续时间信号与离散时间信号。连续时间信号（continuous-time signal）在其所研究的时间内，对任意时刻除若干个不连续点外都有定义。这里"连续"是指函数中的时间取值是连续的，而信号的幅值可连续，也可不连续。常用 $f(t)$ 表示，如图1-1(a)所示。

The independent variables of continuous-time signals are continuous, thus these signals are defined for a continuum of values of the independent variables.

与连续时间信号相对应的是离散时间信号。离散时间信号（discrete-time signal）是指时间（其定义域是一个整数集）是离散的，只在某些不连续的时刻给出函数值，而在其他时间没有定义。常用 $x(n)$ 表示，如图1-1(b)所示。

Discrete-time signals are defined only at discrete times, and consequently, for these signals, the independent variables take on only a discrete set of values.

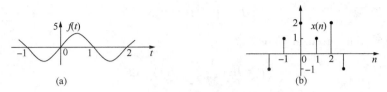

图 1-1　连续时间信号与离散时间信号波形

连续时间信号与离散时间信号的区分要点在于信号的时间是否连续,而不在于幅值是否连续。

对于连续时间信号:幅值连续的称为模拟信号(analog signal);幅值离散的称为脉冲信号(pulse signal)。

对于离散时间信号:幅值连续的称为抽样信号(sampling signal);幅值离散的称为数字信号(digital signal)。

1.2.3　周期信号与非周期信号(Periodic and Aperiodic Signals)

周期信号(periodic signal)是定义在区间$(-\infty, +\infty)$上,且每隔一个固定的时间间隔波形重复变化。连续周期信号与离散周期信号的数学表达式分别为

$$f(t) = f(t+mT) \quad m = \pm 1, \pm 2, \cdots \tag{1-1}$$

$$x(n) = x(n+mN) \quad (n\text{ 取整数}) \tag{1-2}$$

满足以上两式中的最小正数 T 和 N 分别称为周期信号的基本周期。

For a continuous-time signal $f(t)$, if $f(t) = f(t+mT)$ for all values of t, for a discrete-time signal $x(n)$, if $x(n) = x(n+mN)$ for all values of n, in this case, we say that $f(t)[x(n)]$ is periodic with period $T(N)$.

如果两个周期信号的周期之比为有理数(rational number),则它们的和仍然是一个周期信号,其周期为两者周期的最小公倍数(the lowest common multiple),否则为非周期信号。非周期信号(aperiodic signal)就是不具有重复性的信号。

Example 1-1: Determine whether or not each of the following signals is periodic:

① $a\sin t - b\sin 5t$ ② $a\sin t - b\sin \pi t$

Solution: ① For $a\sin t - b\sin 5t$, this is a complex signal, and $T_1 = 2\pi$, $T_2 = 2\pi/5$. In this case, both T_1 and T_2 are rational, the lowest common multiple of T_1 and T_2 is 2π. So, the signal is a periodic signal, and the period is $T = 2\pi$.

② For the signal $a\sin t - b\sin \pi t$, $T_1 = 2\pi$, $T_2 = 2$, $T_1 : T_2 = \pi$, the ratio is an irrational, thus the signal is aperiodic.

在复合信号中,分量周期(或频率)的比为无理数,则该复合信号称为概周期信号,概周期信号是非周期信号。

1.2.4　能量信号与功率信号(Energy and Power Signals)

按照信号的可积性(integrability)划分,信号可以分为能量信号(energy signal)和功率信号(power signal)。

信号可看成是随时间变化的电压或电流,如信号 $f(t)$ 在 1Ω 电阻上的瞬时功率为 $|f(t)|^2$,则在时间区间$(-\infty, +\infty)$所耗的总能量(total energy)为

$$E = \int_{-\infty}^{\infty} |f(t)|^2 dt \qquad (1-3)$$

其平均功率(time-average power)为

$$P = \lim_{T \to \infty} \frac{1}{2T} \int_{-T}^{T} |f(t)|^2 dt \qquad (1-4)$$

能量信号是指能量有限（$0<E<\infty$）的信号。因能量 E 有界，所以当 $T\to\infty$ 时，$P=\lim\limits_{T\to\infty}\dfrac{E}{2T}=0$，表明能量信号平均功率为零。如单个矩形脉冲信号为能量信号，其平均功率为零，只能从能量的角度去考察。

The signals with finite total energy must have zero average power. An example of a finite-energy signal is a signal that takes on the value 1 for $0 \leq t \leq 1$ and 0 otherwise. In this case, $E=1$ and $P=0$.

功率信号是指功率有限（$0<P<\infty$）的信号，因功率 P 有限，故当 $T\to\infty$ 时，$E=\lim\limits_{T\to\infty} 2T \cdot P=\infty$，表明功率信号能量无限。如周期信号、阶跃信号是功率信号，它们的能量为无限，只能从功率的角度去考察。

The signals with finite average power must have infinite energy. Of course this is meaningful, since if there is a nonzero average power per unit time, then integrating this over an infinite time interval yields an infinite amount of energy.

注意，一个信号不可能既是能量信号又是功率信号，但却有少数信号既不是能量信号也不是功率信号。如单位斜坡信号为非功率、非能量信号，它是持续时间无限、幅值也无限的非周期信号。

There are also signals for which neither total energy nor average power are finite. A simple example is the ramp signal.

With these definitions, we can identify three important classes of signals: energy signals, power signals, signals with neither finite energy nor finite power.

1.3 基本的连续时间信号（Basic Continuous-time Signals）

本节介绍几种重要的连续时间信号，这些信号在后续的课程中经常用到。在实际应用中，复杂信号可由这些基本的信号组合而成，并且这些信号对线性系统所产生的响应对分析系统和了解系统的性质起着主导作用，具有普遍意义。

In this section, we introduce several basic continuous-time signals. Not only do these signals occur frequently, but they also serve as basic building blocks from which we can construct many other signals.

1.3.1 实指数信号（Real Exponential Signal）

实指数信号的数学表达式为

$$f(t) = Ce^{at} \qquad (1-5)$$

式中 C 和 a 均为实数，根据 a 的不同取值，有以下三种情况：若 $a>0$，则指数信号幅度随时间增长而增长；若 $a<0$，则指数信号幅度随时间增长而衰减；在 $a=0$ 的特殊情况下，信号不随时间变化，成为直流信号。

Depending upon the values of these parameters, the real exponential signal can exhibit several different characteristics. If $a>0$, it's waveform is growing. If $a<0$, it's waveform is decaying. When $a=0$, it is a direct current signal.

实指数信号的波形如图 1-2 所示。

根据连续时间变量 t 的取值范围，信号 $f(t)$ 又可分为双边信号(bilateral signal)($-\infty<t<\infty$，$t \in \mathbf{R}$)、单边信号(unilateral signal)($-\infty<t<t_1$ 或 $t_2<t<\infty$)和时限信号(time-limited signal)($t_1 \leqslant t \leqslant t_2$)，信号 $f(t)$ 在所对应的区间上有非零确定值。单边信号又分为左边信号(left-sided signal)和右边信号(right-sided signal)。若 $f(t)$ 在 $-\infty<t<t_1$ 区间上为 0，则称为右边信号(若 $t_1=0$，该右边信号称为因果信号(causal signal))；若信号 $f(t)$ 在 $t_2<t<\infty$ 区间上为 0，则称为左边信号。

在实际中遇到较多的因果指数衰减信号(causal decaying exponential signal)，其数学表达式为

$$f(t) = \begin{cases} Ce^{at}, & t \geqslant 0, a < 0 \\ 0, & t < 0 \end{cases} \tag{1-6}$$

波形如图 1-3 所示。

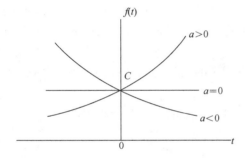

图 1-2　实指数信号

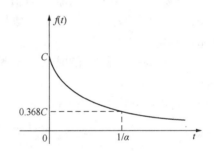

图 1-3　因果指数衰减信号

1.3.2　正弦信号(Sinusoidal Signal)

正弦信号和余弦信号两者仅在相位上相差 $\dfrac{\pi}{2}$，通常统称为正弦信号，其数学表达式为

$$f(t) = K\sin(\omega t + \theta) \tag{1-7}$$

式中 K 为振幅，ω 为角频率，θ 为初始相位。波形如图 1-4 所示。

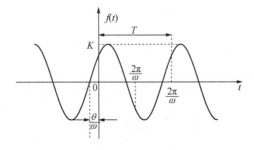

图 1-4　正弦信号

1.3.3 复指数信号(Complex Exponential Signal)

$$f(t) = Ke^{st} \tag{1-8}$$

式中 $s = \sigma + j\omega$ 为复数，K 一般为实数。利用欧拉公式(Euler formula)

$$\begin{cases} e^{j\omega t} = \cos\omega t + j\sin\omega t \\ e^{-j\omega t} = \cos\omega t - j\sin\omega t \end{cases}$$

可将式(1-8)展开，得

$$f(t) = Ke^{st} = Ke^{\sigma t} \cdot e^{+j\omega t} = (Ke^{\sigma t}\cos\omega t) + j(Ke^{\sigma t}\sin\omega t) \tag{1-9}$$

式(1-9)表明，一个复指数信号可分解为实部和虚部两部分。实部(real part)、虚部(imaginary part)都是幅度按指数规律变化的正弦信号。

实指数信号、正弦信号都可由复指数信号导出。根据 σ 和 ω 的不同取值，可得到如下几种信号形式：

$\sigma = 0$，$\omega = 0$ ——直流信号(constant)

$\sigma > 0$，$\omega = 0$ ——升指数信号(growing exponential)

$\sigma < 0$，$\omega = 0$ ——衰减指数信号(decaying exponential)

$\sigma = 0$，$\omega \neq 0$ ——实部和虚部均等幅振荡(the real and imaginary parts are sinusoidal)

$\sigma > 0$，$\omega \neq 0$ ——实部和虚部均增幅振荡(the real and imaginary parts are increasing oscillation)

$\sigma < 0$，$\omega \neq 0$ ——实部和虚部均衰减振荡(the real and imaginary parts are decaying oscillation)

常用的是指数衰减正弦信号(decaying sinusoidal signal)，其数学表达式为

$$f(t) = \begin{cases} Ke^{-at}\sin(\omega t) & t \geq 0 \\ 0 & t < 0 \end{cases}, \quad a > 0$$

波形如图 1-5 所示。

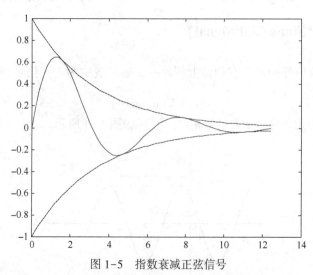

图 1-5 指数衰减正弦信号

1.3.4 抽样信号(Sampling Signal)

Sampling signal plays a veryimportant role in Fourier analysis and in the study of LTI systems. The sampling signal, which is defined as

$$Sa(t) = \frac{\sin t}{t} \tag{1-10}$$

抽样信号具有以下性质：

① $t=0$ 时，借助于罗彼塔法则求得 $Sa(0) = \frac{(\sin t)'}{t'}\big|_{t=0} = \frac{\cos t}{1}\big|_{t=0} = 1$；

② $t \neq 0$ 时，随着 t 的绝对值的增大，函数值的绝对值振荡着不断减小，向 0 趋近；

③ 在 $t = n\pi (n \in Z, n \neq 0)$ 点处，函数值为 0；

④ 该函数是偶函数，即 $Sa(-t) = Sa(t)$；

⑤ $\int_{-\infty}^{\infty} Sa(t) \mathrm{d}t = \pi$，$\int_{-\infty}^{0} Sa(t) \mathrm{d}t = \int_{0}^{\infty} Sa(t) \mathrm{d}t = \frac{\pi}{2}$；

⑥ $\lim\limits_{t \to \pm\infty} Sa(t) = 0$。

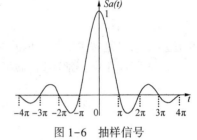

图 1-6 抽样信号

抽样信号的波形如图 1-6 所示。

以相邻两个过零点为端点的区间称为过零区间(zero crossing interval)。由图 1-6 中可以看出原点附近的过零区间宽度为 2π，其他过零区间宽度均为 π。

1.3.5 高斯信号(Gauss Signal)

高斯信号也称钟形脉冲信号，波形如图 1-7 所示，定义为

$$f(t) = E\mathrm{e}^{-(\frac{t}{\tau})^2} \tag{1-11}$$

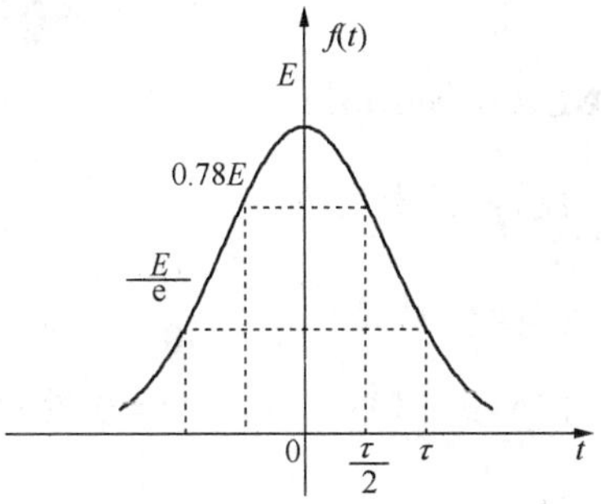

图 1-7 高斯信号

令 $t = \frac{\tau}{2}$，代入函数式求得

$$f\left(\frac{\tau}{2}\right) = E\mathrm{e}^{-\frac{1}{4}} \approx 0.78E \tag{1-12}$$

函数式中的参数 τ 是当 $f(t)$ 由最大值 E 下降为 $0.78E$ 时所占据的时间宽度。高斯信号最重要的性质是其傅里叶变换也是高斯信号，这在信号分析中占有重要地位。

The most important characteristic of the Gauss signal is that its Fourier transform is still Gauss signal, which plays an important role in signal analysis.

下面介绍另一类基本信号——奇异信号(singular signal)，这类信号的数学表达式属于奇

异函数(singular function),即函数本身或其导数或高阶导数出现奇异值(singular value)(趋于无穷)。

1.3.6 单位斜变(斜坡)信号(Unit Ramp Signal)

斜变信号指从某一时刻开始随时间正比例增长的信号,如果增长的变化率为1,则称为单位斜变信号。

The ramp signal, which is a signal with time proportional growth, is an ideal signal. If the growth rate is 1, then it is called the unit ramp signal.

通常用符号 $R(t)$ 表示,其数学表达式为

$$R(t) = \begin{cases} 0, & t<0 \\ t, & t \geqslant 0 \end{cases} \tag{1-13}$$

单位斜变信号是理想信号(ideal signal),是不可实现的。其信号波形如图1-8(a)所示。

在实际应用中,常遇到"截平"的信号,其表达式为

$$R(t) = \begin{cases} \dfrac{k}{\tau}t & (t < \tau) \\ k & (t \geqslant \tau) \end{cases} \tag{1-14}$$

截平信号(truncated signal)波形如图1-8(b)所示。

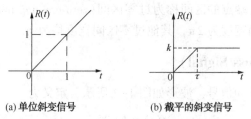

(a) 单位斜变信号　　　(b) 截平的斜变信号

图1-8　斜变信号

1.3.7 单位阶跃信号(Unit Step Signal)

单位阶跃信号以符号 $u(t)$ 表示,其定义为

$$u(t) = \begin{cases} 0 & t<0 \\ 1 & t>0 \end{cases} \tag{1-15}$$

波形如图1-9(a)所示。单位阶跃信号 $u(t)$ 在 $t=0$ 处存在间断点,在此点 $u(t)$ 没有定义。

Obviously, there is a discontinuous point at $t = 0$. As same as the complex exponential signal, the unit step signal will be very important in our examination of the properties of systems.

单位阶跃信号也可以延时任意时刻 t_0,以符号 $u(t-t_0)$ 表示,其表达式为

$$u(t-t_0) = \begin{cases} 0 & t < t_0 \\ 1 & t > t_0 \end{cases} \tag{1-16}$$

对应的波形如图1-9(b)所示。

阶跃信号鲜明地表现出信号的单边特性,即信号在某接入时刻 t_0 以前的幅度为零。利用这一特性可以较方便地以数学表达式描述各种信号的接入特性。例如,用阶跃信号表示矩形脉冲以及对正弦信号的截断,如图1-10所示。其中 $f(t)=u(t)-u(t-t_0)$。

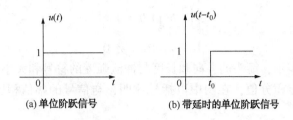

(a) 单位阶跃信号　　　　　(b) 带延时的单位阶跃信号

图 1-9　单位阶跃信号

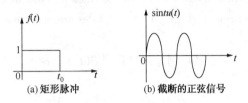

(a) 矩形脉冲　　　　　(b) 截断的正弦信号

图 1-10　截断信号

The relationship between the unit step signal and the unit ramp signal：

$$\frac{dR(t)}{dt} = u(t) \quad \text{—— first derivative} \tag{1-17}$$

$$R(t) = \int_{-\infty}^{t} u(t)\,dt \quad \text{—— running integral} \tag{1-18}$$

The unit ramp signal is also represented as $R(t) = tu(t)$.

1.3.8　符号函数(Sign Signal)

符号函数的数学表达式为

$$\mathrm{sgn}(t) = \begin{cases} 1, & t > 0 \\ -1, & t < 0 \end{cases} \tag{1-19}$$

与阶跃信号类似，符号函数在跳变点可不予定义。其波形如图 1-11 所示。

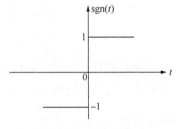

图 1-11　符号函数

显然，阶跃信号可用来表示符号函数，即

$$\mathrm{sgn}(t) = 2u(t) - 1 \text{ 或 } \mathrm{sgn}(t) = u(t) - u(-t) \tag{1-20}$$

1.3.9　单位冲激信号(Unit Impulse Signal)

某些物理现象需要用一个持续时间无穷短而取值无穷大，但对时间的积分值为有限值的函数模型来描述。例如，电学中的雷击电闪，数字通信中的抽样脉冲，力学中瞬间作用的冲击力等。

The unit impulse function is quite useful in the analysis of the signals and systems. The unit impulse signal is denoted as $\delta(t)$ and can be defined on many ways.

(1) 冲激信号的定义(Definition of the Impulse Signal)

冲激信号可以有不同的定义方式，例如，由矩形脉冲、三角形脉冲演变为冲激函数，还可利用指数函数、钟形函数、抽样函数、狄拉克(Dirac)函数来定义。

单位冲激信号的狄拉克(Dirac)定义为

$$\begin{cases} \int_{-\infty}^{\infty} \delta(t) \, dt = 1 \\ \delta(t) = 0, \ t \neq 0 \end{cases} \tag{1-21}$$

冲激信号用箭头表示，箭头的方向和长度与冲激强度的符号和大小一致，其冲激强度就是冲激信号对时间的定积分值，在图中以括号注明，与信号的幅值相区分，如图 1-12(a) 所示。

Since $\delta(t)$ has, in effect, no duration but unit area, we adopt the graphical notation for it shown in Figure 1-12(a), where the arrow at $t=0$ indicates that the area of the pulse is concentrated at $t=0$ and the height of the arrow and the "1" next to the arrow are used to represented the area of the impulse.

单位冲激信号可以延时至任意时刻 t_0，以符号 $\delta(t-t_0)$ 表示。冲激点在 t_0、冲激强度为 E 的冲激信号 δ_{E,t_0} 定义为

$$\begin{cases} \int_{-\infty}^{\infty} \delta_{E,t_0}(t) \, dt = E \\ \delta_{E,t_0}(t) = 0, \ (t \neq t_0) \end{cases} \tag{1-22}$$

波形如图 1-12(b) 所示。

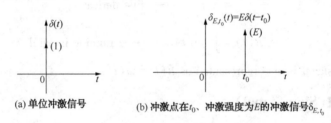

(a) 单位冲激信号　　(b) 冲激点在 t_0、冲激强度为 E 的冲激信号 δ_{E,t_0}

图 1-12　冲激信号

冲激信号是作用时间极短，但取值极大的一类信号的数学模型。例如，单位阶跃信号加在不含初始储能的电容两端，t 从 0^- 到 0^+ 的极短时刻，电容两端的电压从 0V 跳变到 1V，而流过电容的电流 $i(t) = \dfrac{C du(t)}{dt}$ 为无穷大，这种电流持续时间为零，电流幅度为无穷大，但电流的时间积分有限的物理现象就可以用冲激函数 $\delta(t)$ 来描述。

为了更为直观地理解冲激信号，我们还可以将其看成某些普通信号的极限。例如，由对矩形脉冲取极限表示的单位冲激函数为

$$\delta(t) = \lim_{\tau \to 0} \frac{1}{\tau} \left[u\left(t + \frac{\tau}{2}\right) - u\left(t - \frac{\tau}{2}\right) \right] \tag{1-23}$$

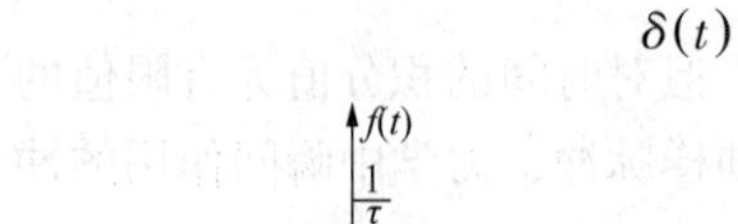

如图 1-13 所示，宽度为 τ，高度为 $\dfrac{1}{\tau}$ 的矩形脉冲，当保持矩形脉冲的面积 $\tau \cdot \dfrac{1}{\tau} = 1$ 不变，而使脉宽 τ 趋于零时，脉高 $\dfrac{1}{\tau}$ 必为无穷大，此极限情况即为单位冲激信号。

图 1-13　矩形脉冲的极限模型

$\delta(t)$ is a short pulse, of duration τ and with unit area value of τ. As $\tau \to 0$, $\delta(t)$ becomes narrower and higher, maintaining its unit area. Its limiting form can then be thought of as an idealization of the short pulse $\delta(t)$ as the duration τ becomes insignificant.

The relationship between the unit step signal and the unitimpulse signal:

$$u(t) = \int_{-\infty}^{t} \delta(\tau) d\tau \qquad (1-24)$$

$$\delta(t) = \frac{du(t)}{dt} \qquad (1-25)$$

That is, $u(t)$ is the running integral of the unit impulse signal. This suggests that the continuous-time unit impulse signal can be thought of as the first derivative of $u(t)$.

(2) 冲激信号的性质(Properties of the Impulse Signal)

① 抽样特性(sampling property)

原点抽样:
$$\int_{-\infty}^{\infty} f(t)\delta(t) dt = f(0) \qquad (1-26)$$

延迟抽样:
$$\int_{-\infty}^{\infty} f(t)\delta(t-t_0) dt = f(t_0) \qquad (1-27)$$

满足抽样特性的前提条件是 $f(t)$ 在抽样点处连续。

Proof: We can deduce the property from the definition of the unit impulse signal,

$$\int_{-\infty}^{\infty} f(t)\delta(t) dt = \int_{-\infty}^{0^-} f(t)\delta(t) dt + \int_{0^-}^{0^+} f(t)\delta(t) dt + \int_{0^+}^{\infty} f(t)\delta(t) dt$$

$$= 0 + \int_{0^-}^{0^+} f(t)\delta(t) dt + 0$$

$$= f(0) \int_{0^-}^{0^+} \delta(t) dt$$

$$= f(0)$$

原点抽样性质表明一个在原点连续的信号与冲激信号相乘以后的积分就等于该信号在原点处的值。

同理，我们也可以证明延迟抽样性质。延迟抽样性质表明一个在 t_0 点处连续的信号与冲激信号相乘以后的积分就等于该信号在 t_0 点处的值。

从冲激信号的抽样特性，我们还可以得出

$$f(t)\delta(t) = f(0)\delta(t) \qquad (1-28)$$

$$f(t-t_0)\delta(t-t_1) = f(t_1-t_0)\delta(t-t_1) \qquad (1-29)$$

$$\int_{-\infty}^{\infty} f(t-t_0)\delta(t-t_1) dt = f(t_1-t_0) \qquad (1-30)$$

② 对称性(symmetry property)

$$\delta(t) = \delta(-t) \qquad (1-31)$$

Proof: Because $\int_{-\infty}^{\infty} \delta(-t) dt = -\int_{\infty}^{-\infty} \delta(x) dx = \int_{-\infty}^{\infty} \delta(x) dx = \int_{-\infty}^{\infty} \delta(t) dt$

hence, $\delta(t) = \delta(-t)$.

③ 尺度特性(scale property)

$$\delta(at) = \frac{1}{|a|}\delta(t) \qquad (1-32)$$

Proof: When $a>0$, the left hand side of above equation is equal to

$$\int_{-\infty}^{\infty} \delta(at)\,dt = \frac{1}{a}\int_{-\infty}^{\infty} \delta(x)\,dx = \frac{1}{a}$$

When $a<0$, the left hand side is equal to

$$\int_{-\infty}^{\infty} \delta(at)\,dt = \frac{1}{a}\int_{+\infty}^{-\infty} \delta(x)\,dx = \frac{-1}{|a|}\int_{+\infty}^{-\infty} \delta(x)\,dx = \frac{1}{|a|}$$

The right hand side, then, is equal to

$$\int_{-\infty}^{\infty} \frac{1}{|a|}\delta(t)\,dt = \frac{1}{|a|}\int_{-\infty}^{\infty} \delta(t)\,dt = \frac{1}{|a|}$$

Therefore, the result of left side equals to that of the right side, that's all.

Example 1-2: Calculate the following integral according to the properties of impulse signal.

① $\int_{-\infty}^{\infty} f(t-t_0)\delta(t)\,dt$ ② $\int_{-\infty}^{\infty} (t+\sin t)\delta(t-\frac{\pi}{6})\,dt$

③ $\int_{-\infty}^{\infty} (t+1)^2\delta(-2t)\,dt$ ④ $\int_{-\infty}^{\infty} (t+1)^2\delta(1-2t)\,dt$

Solution:

① $\int_{-\infty}^{\infty} f(t-t_0)\delta(t)\,dt = f(0-t_0) = f(-t_0)$

② $\int_{-\infty}^{\infty} (t+\sin t)\delta(t-\frac{\pi}{6})\,dt = \frac{\pi}{6} + \sin\left(\frac{\pi}{6}\right) = \frac{\pi}{6} + \frac{1}{2}$

③ $\int_{-\infty}^{\infty} (t+1)^2\delta(-2t)\,dt = \int_{-\infty}^{\infty} (0+1)^2 \frac{1}{2}\delta(t)\,dt = \frac{1}{2}$

④ $\int_{-\infty}^{\infty} (t+1)^2\delta(1-2t)\,dt = \int_{-\infty}^{\infty} (t+1)^2 \frac{1}{2}\delta(t-\frac{1}{2})\,dt = \frac{1}{2}\left(\frac{1}{2}+1\right)^2 = \frac{9}{8}$

1.5 基本的离散时间信号（Basic Discrete-time Signals——sequence）

时间为离散变量的信号称为离散时间信号，它只在离散时间上给出函数值，是时间上不连续的序列，常用 $x(n)$ 表示。在离散时域中，有一些基本的离散时间信号，它们在离散信号与系统中起着重要的作用，有些信号和前面讨论的连续时间信号相似，但也有不同之处，下面给出一些典型的离散信号表达式和波形。

Discrete-time signals are defined only at discrete times, and consequently, for these signals, the independent variable takes on only a discrete set of values. A discrete-time signal $x(n)$ may represent a phenomenon for which the independent variable is inherently discrete.

1.4.1 单位脉冲序列(Unit Impulse Sequence)

单位脉冲序列用符号 $\delta(n)$ 表示，定义如下：

$$\delta(n) = \begin{cases} 1, & n=0 \\ 0, & n \neq 0 \end{cases} \tag{1-33}$$

其波形如图 1-14 所示。

$\delta(n)$ 在 $n=0$ 时有确定值 1，这与 $\delta(t)$ 在 $t=0$ 时的情况不同。有位移的单位脉冲序列表达式如下所示：

$$\delta(n-m) = \begin{cases} 1, & n = m \\ 0, & n \neq m \end{cases}$$

单位脉冲序列在离散信号与系统的分析、综合中有着重要的作用,其地位犹如连续时间信号与系统中的单位冲激信号 $\delta(t)$。在实际中,$\delta(t)$ 是不存在的,而 $\delta(n)$ 是存在的。

任意序列可以表示成单位脉冲序列的移位加权和,即

$$x(n) = \sum_{m=-\infty}^{\infty} x(m)\delta(n-m) \tag{1-34}$$

1.4.2 单位阶跃序列(Unit Step Sequence)

单位阶跃序列 $u(n)$ 如图 1-15 所示,定义如下:

$$u(n) = \begin{cases} 1, & n \geq 0 \\ 0, & n < 0 \end{cases} \tag{1-35}$$

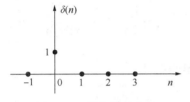

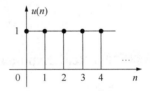

图 1-14 单位脉冲序列　　　　图 1-15 单位阶跃序列

它类似于连续时间信号与系统的单位阶跃信号 $u(t)$。但 $u(t)$ 在 $t=0$ 时常不给予定义,而 $u(n)$ 在 $n=0$ 时定义为 $u(0)=1$。观察 $\delta(n)$ 序列与 $u(n)$ 序列的定义式,可以看出两者之间的关系为

$$u(n) = \delta(n) + \delta(n-1) + \delta(n-2) + \delta(n-3) + \cdots = \sum_{k=0}^{\infty} \delta(n-k) \tag{1-36}$$

$$\delta(n) = u(n) - u(n-1) \tag{1-37}$$

There is a close relationship between the unit impulse sequence and unit step sequence. In particular, the unit impulse sequence is the first difference of the unit step sequence, and the unit step sequence is the running sum of the unit impulse sequence.

The unit impulse sequence can be used to sample the value of a signal at $n=0$. In particular, since $\delta(n)$ is nonzero (and equal to 1) only for $n=0$, it follows that

$$x(n)\delta(n) = x(0)\delta(n) \tag{1-38}$$

more generally, if we consider a unit impulse $\delta(n-n_0)$ at $n=n_0$, then

$$x(n)\delta(n-n_0) = x(n_0)\delta(n-n_0) \tag{1-39}$$

this sampling property of the unit impulse will play an important role in later chapter.

1.4.3 单位矩形序列(Unit Rectangular Sequence)

单位矩形序列用 $R_N(n)$ 表示,定义为

$$R_N(n) = \begin{cases} 1, & 0 \leq n \leq N-1 \\ 0, & n < 0, n \geq N \end{cases} \tag{1-40}$$

亦可用 $\delta(n)$、$u(n)$ 表示 $R_N(n)$,即

$$R_N(n) = \delta(n) + \delta(n-1) + \delta(n-2) + \cdots + \delta(n-N+1) = \sum_{m=0}^{N-1} \delta(n-m) \tag{1-41}$$

$$R_N(n) = u(n) - u(n-N) \tag{1-42}$$

单位矩形序列的波形如图 1-16 所示。

1.4.4 斜变序列(Unit Ramp Sequence)

斜变序列是包络为线性变化的序列，表达式为

$$x(n) = nu(n) \tag{1-43}$$

波形如图 1-17 所示。

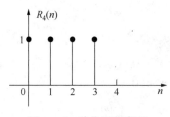
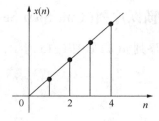

图 1-16　单位矩形序列　　　　　图 1-17　斜变序列

1.4.5 实指数序列(Real Exponential Sequence)

实指数序列表达式为

$$x(n) = a^n u(n) \tag{1-44}$$

波形如图 1-18 所示。

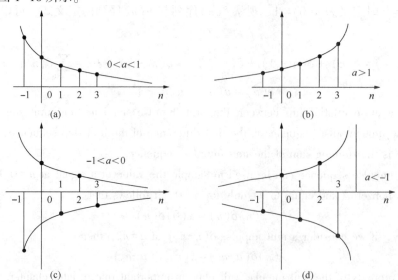

图 1-18　实指数序列

其中 a 为实数，其波形特点是：当 $|a|>1$ 时，序列发散；当 $|a|<1$ 时，序列收敛；当 a 为负数时，序列值正负交替出现。

1.4.6 正弦型序列(Sinusoidal Sequence)

正弦型序列是包络为正、余弦变化的序列。例如，$\sin n\theta_0$，$\cos n\theta_0$，若 $\theta_0 = \dfrac{\pi}{5}$，$N = \dfrac{2\pi}{\dfrac{\pi}{5}} = 10$，

即每10个点重复一次正、余弦变化，如图1-19所示。

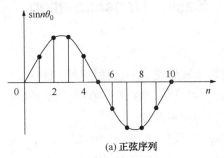
(a) 正弦序列

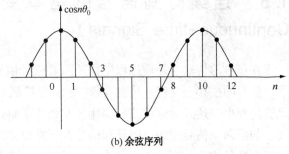
(b) 余弦序列

图1-19 正、余弦序列

正弦型序列一般表示为
$$x(n) = A\sin(n\theta_0 + \varphi_n) \tag{1-45}$$

1.4.7 周期序列(Periodic Sequence)

$$x(n) = x(n+N), \quad -\infty < n < \infty \tag{1-46}$$

则该序列为周期序列，周期为 N 点。

对模拟周期信号采样得到的序列，未必是周期序列。例如，模拟正弦型采样信号一般表示为
$$x(n) = A\cos(n\theta_0 + \varphi_n) = A\cos(n\theta_0 + \varphi_n)$$

式中，$\dfrac{2\pi}{\theta_0} = \dfrac{2\pi}{\omega_0 T} = \dfrac{2\pi f_s}{\omega_0} = \dfrac{f_s}{f_0}$，$f_s$ 为采样频率(sampling frequency)，f_0 为模拟周期信号频率。

可由以下条件判断 $x(n)$ 是否为周期序列：

① 若 $\dfrac{2\pi}{\theta_0} = N$，$N$ 为整数，则 $x(n)$ 是周期序列，且周期为 N。

If $\dfrac{2\pi}{\theta_0} = N$ is an integer, then the fundamental period is N. For example, $x(n) = \sin n\theta_0$, if $\theta_0 = \dfrac{\pi}{5}$, then the period $N = \dfrac{2\pi}{\dfrac{\pi}{5}} = 10$.

② 若 $\dfrac{2\pi}{\theta_0} = S = \dfrac{N}{L}$，$L$、$N$ 为整数，则 $x(n)$ 是周期序列，且周期为 $N = SL$。

If $\dfrac{2\pi}{\theta_0} = S = \dfrac{N}{L}$ is a rational number, then the fundamental period is $N = SL$. For example, $x(n) = \sin n\theta_0$, if $\theta_0 = \dfrac{8\pi}{3}$, then the period $N = 3$.

③ 若 $\dfrac{2\pi}{\theta_0}$ 为无理数，则 $x(n) = A\cos(n\theta_0 + \varphi_n)$ 不是周期序列。

If $\dfrac{2\pi}{\theta_0}$ is an irrational number, then the sinusoidal sequence will not be periodic at all. For example, $x(n) = \sin n\theta_0$, if $\theta_0 = \dfrac{1}{4}$, then $\dfrac{2\pi}{\theta_0} = 8\pi$, is an irrational number, so $x(n)$ is not periodic sequence.

1.5 连续时间信号的基本变换（Basic Transformation of Continuous-time Signals）

在信号的传输与处理过程中往往需要进行信号的运算及波形变换，包括信号的相加或相乘，信号的位移、反转、尺度变换（压缩与扩展）、微分、积分等。在这一节中，需要熟悉运算过程中表达式对应的波形变化，并初步了解这些运算的物理背景。

如果两个信号相加，则其和信号在任意时刻的幅值等于两信号在该时刻的幅值之和。假如两个信号相乘，则其积信号在任意时刻的幅值等于两信号在该时刻的幅值之积。

A central concept in signal and system analysis is that of the transformation of a signal. In this section, we focus on a very limited but important class of elementary signal transformations that involve simple modification of the independent variable, i.e., the time axis. As we will see in this section, these elementary transformations allow us to introduce several basic properties of signals and systems.

1.5.1 信号的时移（Time Shifting）

信号的时移也称信号的位移、时延。将信号 $f(t)$ 的自变量 t 用 $t-t_0$ 替换，得到的信号 $f(t-t_0)$ 就是 $f(t)$ 的时移，它是 $f(t)$ 的波形在时间 t 轴上整体移位 t_0。

If $t_0 > 0$, then the waveform of $f(t-t_0)$ is obtained by shifting $f(t)$ toward the right, relative to the time axis. If $t_0 < 0$, then the waveform of $f(t)$ is shifted to the left. The transformation waveform of $f(t-t_0)$ is shown in the Figure 1-20.

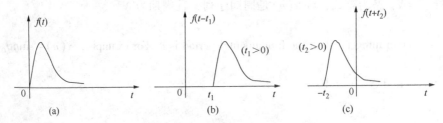

Figure 1-20　time shifting of signal

1.5.2 信号的反转（Reflection of Signal）

将信号的自变量 t 换成 $-t$，可以得到另一个信号 $f(-t)$，这种变换称之为信号的反转，或称之为反褶、折叠。它的几何意义是将自变量轴"倒置"，取其原信号自变量轴的负方向作为变换后信号自变量轴的正方向。

The signal $f(-t)$ represents a reflection version of $f(t)$ about $t = 0$, it's transformation waveform is shown in the Figure 1-21.

1.5.3 信号的尺度变换（Scaling）

将 $f(t)$ 的自变量 t 用 $at(a>0)$ 替换，得到 $f(at)$ 称为 $f(t)$ 的尺度变换，其波形是 $f(t)$ 波形在时间轴上的压缩或扩展。

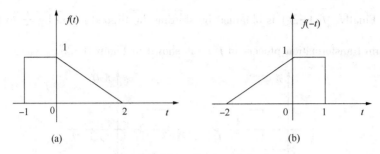

Figure 1-21　reflection of signal

If $a>1$, the signal $f(at)$ is a compressed version of $f(t)$. If $0<a<1$, the signal $f(at)$ is an expanded (stretched) version of $f(t)$.

例如，假设$f(t)=\sin\omega_0 t$是正常语速的信号，则$f(2t)=\sin 2\omega_0 t=f_1(t)$是2倍语速的信号，而$f(\frac{t}{2})=\sin(\frac{1}{2}\omega_0 t)=f_2(t)$是降低一半语速的信号。$f_1(t)$与$f_2(t)$在时间轴上被压缩或扩展，但幅度均没有变化，如图1-22所示。

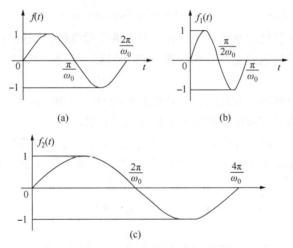

图 1-22　信号的尺度变换

上面对信号的时移、反转和尺度变换分别进行了描述。实际上，信号的变化常常是上述三种方式的综合，即信号$f(t)$变化为$f(at+b)$（其中$a\neq 0$）。现举例说明其变化过程。

It is often of interest to determine the effect of transforming the independent variable of a given signal $f(t)$ to obtain a signal of the form $f(at+b)$, where a and b are given numbers. Such a transformation of the independent variable preserves the shape of $f(t)$, except that the resulting signal may be linearly stretched if $|a|<1$, linearly compressed if $|a|>1$, reversed in time if $a<0$, and shifted in time if b is nonzero. This is illustrated in the following example.

例 1-3：已知信号$f(t)$的波形如图1-23(a)所示，试画出$f(1-2t)$的波形。

解：一般说来，在t轴尺度保持不变的情况下，信号$f(at+b)$（$a\neq 0$）的波形可以通过对信号$f(t)$波形的时移、反转(若$a<0$)和尺度变换得到。根据变换操作顺序不同，可用多种方法画出$f(1-2t)$的波形。

这里我们按"反转→尺度→时移"顺序求解。

Firstly, let's flip $f(t)$ to obtain $f(-t)$, and $f(-2t)$ is obtained by linearly compression $f(-t)$

by a factor of 2. Finally, $f(1-2t)$ is obtained by shifting the flipped signal by $\dfrac{1}{2}$ to the right.

The waveform transformation process of $f(t)$ is shown in Figure 1-23.

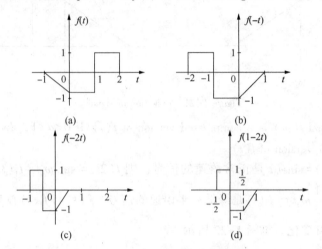

Figure 1-23　the waveform transformation process of $f(t)$

信号的时移、反转和尺度变换只是函数自变量的简单变换,变换结果与变换的操作顺序无关。读者可自行练习其他的变换操作顺序,如"时移→反转→尺度"的顺序,验证变换结果是否相同。

从上面的分析可以看出,变换前后信号端点的函数值不变,因此,可以通过端点函数值不变这一关系来确定信号变换前后其图形中各端点的位置。

设变换前的信号为 $f(t)$,变换后为 $f(at+b)$,t_1 与 t_2 对应变换前信号 $f(t)$ 的左右端点坐标,t_{11} 与 t_{22} 对应变换后信号 $f(at+b)$ 的左右端点坐标。由于信号变化前后端点的函数值不变,故有

$$\begin{cases} f(t_1) = f(at_{11} + b) \\ f(t_2) = f(at_{22} + b) \end{cases} \tag{1-47}$$

根据上述关系可以求解出变换后信号的左右端点坐标 t_{11} 和 t_{22},即

$$\begin{cases} t_1 = at_{11} + b \\ t_2 = at_{22} + b \end{cases} \Rightarrow \begin{cases} t_{11} = \dfrac{1}{a}(t_1 - b) \\ t_{22} = \dfrac{1}{a}(t_2 - b) \end{cases} \tag{1-48}$$

上述方法过程简单,特别适合信号从 $f(mt+n)$ 变换到 $f(at+b)$ 的过程。因为此时若按原先的方法,需将信号 $f(mt+n)$ 经过平移、反转、再展缩的逆过程得到信号 $f(t)$,再将信号 $f(t)$ 经过反转、展缩、再平移的过程得到信号 $f(at+b)$。若根据信号变换前后的端点函数值不变的原理,则可以很简便地计算出变换后信号的端点坐标,从而得到变换后的信号 $f(at+b)$,其计算公式如下:

$$\begin{cases} f(mt_1 + n) = f(at_{11} + b) \\ f(mt_2 + n) = f(at_{22} + b) \end{cases} \tag{1-49}$$

根据上述关系可以求解出变换后信号的左右端点坐标 t_{11} 和 t_{22},即

$$\begin{cases} mt_1 + n = at_{11} + b \\ mt_2 + n = at_{22} + b \end{cases} \Rightarrow \begin{cases} t_{11} = \dfrac{1}{a}(mt_1 + n - b) \\ t_{22} = \dfrac{1}{a}(mt_2 + n - b) \end{cases} \quad (1-50)$$

Example 1-4: According to the principle of the endpoint function values are unchanged before and after the transformation to solve the case of Example 1-3.

Solution: From $f(t)$ to $f(1-2t)$, we have $t_1 = -1$, $t_2 = 2$, $a = -2$, $b = 1$, $t_{11} = 1$, $t_{22} = -\dfrac{1}{2}$, that is, the endpoint value $t_1 = -1$ of $f(t)$ corresponds to the endpoint value $t_{11} = 1$ of $f(1-2t)$, and the endpoint value $t_2 = 2$ of $f(t)$ corresponds to the endpoint value $t_{22} = -\dfrac{1}{2}$ of $f(1-2t)$. Therefore, the waveform after transformation is shown in Figure 1-24. Obviously, the waveform is the same as that of Example 1-3.

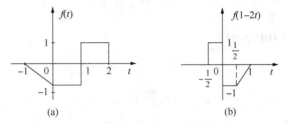

Figure 1-24 the waveform of the Example 1-4

Example 1-5: Given the waveform of signal $f(2t+3)$, shown in Figure 1-25(a), try to plot the waveform of $f(-3t+6)$.

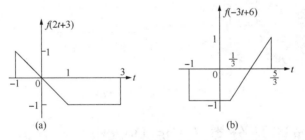

Figure 1-25 the waveform of the Example 1-5

Solution: From $f(2t+3)$ to $f(-3t+6)$, we have $t_1 = 1$, $t_2 = 3$, $m = 2$, $n = 3$, $a = -3$, $b = 6$, according to Eq. (1-49) yields

$$t_{11} = -\dfrac{1}{3}[2(-1)+3-6] = \dfrac{5}{3}$$

$$t_{22} = -\dfrac{1}{3}(2 \times 3 + 3 - 6) = -1$$

that is, the endpoint value $t_1 = 1$ of $f(2t+3)$ corresponds to the endpoint value $t_{11} = \dfrac{5}{3}$ of $f(-3t+6)$, and the endpoint value $t_2 = 3$ of $f(t)$ corresponds to the endpoint value $t_{22} = -1$ of $f(-3t+6)$. The waveform of $f(-3t+6)$ is illustrated in Figure 1-25(b).

1.5.4 信号的微分（Differentiation）

信号的微分是指信号对时间的导数（derivative），可表示为

$$f'(t) = \frac{df(t)}{dt} \tag{1-51}$$

信号经过微分后突出了变化部分。若信号有间断点（discontinuity point），间断点处的微分可用冲激函数表示，如图1-26所示。

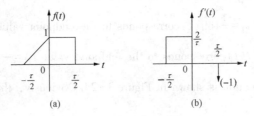

图1-26 信号的微分

1.5.5 信号的积分（Integral）

一个连续信号的积分是指信号在区间$(-\infty, t)$上的积分（running integral），可表示为

$$y(t) = \int_{-\infty}^{t} f(\tau) d\tau \tag{1-52}$$

也就是求从$-\infty$到任一瞬间t，曲线$f(\tau)$下面覆盖的面积。信号经过积分后平滑了变化部分，信号的积分如图1-27所示。

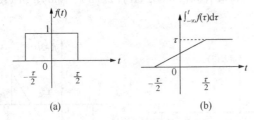

图1-27 信号的积分

1.6 系统的描述及分类（The Description and Classification of Systems）

要产生信号，并对信号进行传输、处理、存储和转化，需要一定的物理装置，即系统。如通信系统、控制系统、经济系统、生态系统等，其中，电系统是应用最为广泛的系统之一。

通常，只有单个输入和单个输出信号的系统称为单输入单输出系统。如果含有多个输入、输出信号，则称为多输入多输出系统。

Physical systems in the broadest sense are an interconnection of components, devices, or subsystems. In contexts ranging from signal processing and communications to electromechanical motors, automotive vehicles, and chemical-processing plants, a system can be viewed as a process in which input signals are transformed by the system or cause the system to respond in some way, resulting in other signals an outputs.

1.6.1 系统模型(System Model)

系统模型是系统物理特性的数学抽象,以数学表达式或具有理想特性的符号组合图形来表征系统特性。模型(model)并非物理实体,它由一些理想元件组合而成,每个理想元件各代表着系统的一种特性,这些理想元件的连接不必与系统中实际元件的组成构成完全相当,但它们结合的总体所呈现的特性与实际系统的特性相近。

The mathematical descriptions of systems from a wide variety of applications frequently have a great deal in common, and it is this fact that provides considerable motivation for the development of broadly applicable tools for signal and system analysis.

(1) 数学表达式描述数学模型(using the mathematical expressions to describe the mathematical model)

在建立系统模型方面,系统的数学描述方法可以分为两大类:一类是输入-输出描述法(input - output description);另一类是状态变量分析法(state variable analysis method)。

① 输入-输出描述法(input - output description)

输入-输出描述法着眼于系统激励(excitation signal/input signal)与响应(response)之间的关系,并不关心系统内部变量(internal variables)的情况。通常,连续时间系统用微分方程(differential equation)描述,离散时间系统用差分方程(difference equation)描述。

② 状态变量分析法(state variable analysis method)

状态变量分析法就是把系统内独立的物理变量作为状态变量(state variable),利用状态变量与输入变量(input variable)、状态变量与输出变量(output variable)描述系统特性的方法。状态变量分析法特别适用于多输入-多输出系统。在控制系统的理论研究中,广泛采用状态变量分析法。

(2) 框图表示系统模型(using the block diagram to describe the system model)

除利用数学表达式描述系统模型之外,也可借助框图(the block diagram)表示系统模型,每个框图反映某种数学运算功能,如果给出每个框图输出与输入信号的约束条件(constraint sondition),那么若干个框图就可以组成一个完整的系统。利用线性微分(差分)方程的基本运算单元(operation units)给出系统框图的方法也称为系统仿真(simulation)(或模拟)。后续课程中将详细介绍利用基本运算单元构成多种多样的框图及其组合形式来表示系统。

1.6.2 系统分类(The Classification of Systems)

在信号与系统分析中,常以系统的数学模型和基本特性分类。系统可分为连续时间系统与离散时间系统;线性系统与非线性系统;时变系统与非时变系统;因果系统与非因果系统;稳定系统与非稳定系统等。

(1) 连续时间系统与离散时间系统(continuous-time systems and discrete-time systems)

如果一个系统输入信号和输出信号均为连续时间信号,则该系统称为连续时间系统。同样,如果一个系统输入信号和输出信号均为离散时间信号,则该系统称为离散时间系统。

A continuous-time system is one in which continuous-time input signals are transformed into continuous - time output signals. Similarly, a discrete - time system, that is, a system that transforms discrete-time inputs into discrete-time outputs.

一般情况下，连续时间系统只能处理连续时间信号，离散时间系统只能处理离散时间信号。但在引入某些信号变换的部件后，就可以使连续时间系统处理离散时间信号，离散时间系统处理连续时间信号。例如，连续时间信号经过 A/D 转换器（A/D converter）后就可以由离散时间系统处理。连续时间系统与离散时间系统通常采用图 1-28 所示符号表示。

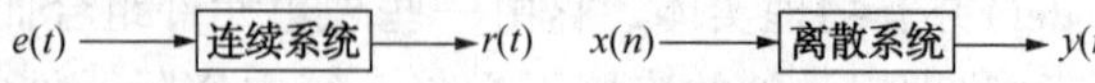

$e(t)$ ⟶ 连续系统 ⟶ $r(t)$ $x(n)$ ⟶ 离散系统 ⟶ $y(n)$

图 1-28　连续时间系统与离散时间系统的符号表示

（2）线性系统与非线性系统（linear and nonlinear systems）

线性系统是指具有线性特性的系统。线性特性包括齐次性和叠加性。所谓齐次性是指当输入信号乘以某常数时，响应也乘以相同的常数；而叠加性是指当几个激励信号同时作用于系统时，总的输出响应等于每个激励单独作用所产生的响应之和。同时具有叠加性和齐次性的系统才称之为线性系统，否则为非线性系统。

下面用公式的形式来表示系统的线性特性。

齐次性：若 $e(t) \to r(t)$，则 $ke(t) \to kr(t)$；

叠加性：若 $e_1(t) \to r_1(t)$，$e_2(t) \to r_2(t)$，则 $e_1(t) + e_2(t) \to r_1(t) + r_2(t)$；

线性：若 $e_1(t) \to r_1(t)$，$e_2(t) \to r_2(t)$，则 $k_1 e_1(t) + k_2 e_2(t) \to k_1 r_1(t) + k_2 r_2(t)$。

A linear system is a system that possesses the important property of superposition: if an input consists of the weighted sum of several signals, then the output is the superposition—that is, the weighted sum—of the responses of the system to each of those signals. More precisely, let $r_1(t)$ be the response of a continuous-time system to an input $e_1(t)$, and let $r_2(t)$ be the output corresponding to the input $e_2(t)$. Then the system is linear if:

① The response to $e_1(t) + e_2(t)$ is $r_1(t) + r_2(t)$.

② The response to $ke_1(t)$ is $kr_1(t)$, where k is any complex constant.

The first of these two properties is known as the additivity property, the second is known as the scaling or homogeneity property. The two properties defining a linear system can be combined into a single statement: $k_1 e_1(t) + k_2 e_2(t) \to k_1 r_1(t) + k_2 r_2(t)$.

由常系数微分方程描述的系统，如果初始状态为零，则系统满足叠加性与齐次性。若初始状态不为零，必须将外加激励信号与初始状态的作用分别处理才能满足叠加性与齐次性，否则容易引起混淆。通常，以线性微分方程作为输入、输出描述方程的系统都是线性系统，而以非线性微分方程作为输入、输出描述方程的系统都是非线性系统。

实际上许多连续时间系统和离散时间系统都含有初始状态（initial state）。对于具有初始状态的线性系统，输出响应等于零输入响应 $r_{zi}(t)$（zero input response）与零状态响应 $r_{zs}(t)$（zero state response）之和。因此，在判断具有初始状态的系统是否线性时，应从三个方面来判断。①分解性：即系统的输出响应等于零输入响应与零状态响应之和；②零输入线性：即系统的零输入响应必须对所有的初始状态呈现线性特性；③零状态线性：即系统的零状态响应必须对所有的输入信号呈现线性特性。只有三个条件都符合，该系统才为线性系统。

Example 1-6: Determine whether or not each of the following systems is linear.

① $r(t) = \int_{-\infty}^{t} e(\tau) d\tau$

② $r(t) = ae(t) + b$, a, b are constant. If $e(0^-) = b$, then is the system linear?

Solution: ① Let $e(t)=k_1e_1(t)+k_2e_2(t)$, then
$$r(t) = \int_{-\infty}^{t} [k_1e_1(\tau) + k_2e_2(\tau)]d\tau$$
$$= k_1\int_{-\infty}^{t} e_1(\tau)d\tau + k_2\int_{-\infty}^{t} e_2(\tau)d\tau = k_1r_1(t) + k_2r_2(t)$$

therefore, the system is linear.

② If $e(0^-)=0$, then $a[ke(t)]+b \neq k[ae(t)+b]$, so the system is nonlinear, if $e(0^-)=b$, then $r(t)=ae(t)+e(0^-)=r_{zs}(t)+r_{zi}(t)$, satisfies the linear property, so, the system is linear when $e(0^-)=b$.

(3) 时不变系统与时变系统(time invariant systems and time-varying systems)

参数不随时间变化的系统,称为时不变系统,否则称为时变系统。一个时不变系统,由于参数不随时间变化,故系统的输入、输出关系也不会随时间变化。

A system is time invariant if the behavior and characteristics if the system are fixed over time. The property of time invariance can be described very simply in terms of the signals and systems language that we have introduced. Specifically, a system is time invariant if a time shift in the input signal results in an identical time shift in the output signal.

如果激励 $e(t)$ 作用于系统产生的零状态响应为 $r(t)$,那么,当激励延迟 t_d 接入时,其零状态响应也延迟相同的时间,其响应的波形形状保持相同。用公式表示如下:
$$e(t) \rightarrow r(t)$$
$$e(t-t_d) \rightarrow r(t-t_d)$$

时不变连续系统示意图如图 1-29 所示。

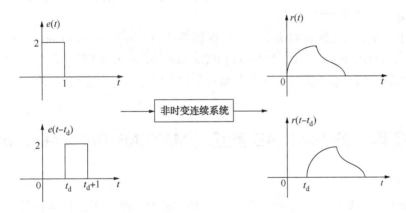

图 1-29 时不变连续时间系统示意图

Example 1-7: Determine whether or not each of the following systems is time invariant.

① $r(t) = \int_{-\infty}^{t} e(\tau)d\tau$ ② $r(t) = \sin t \cdot e(t)$

Solution: ① Let $e(t) \rightarrow e(t-t_d)$, then
$$\int_{-\infty}^{t} e(\tau-t_d)d\tau \xrightarrow{\tau-t_d=\lambda} \int_{-\infty}^{t-t_d} e(\lambda)d\lambda = r(t-t_d)$$

So that the system is time invariant.

② Because $\sin t \cdot e(t-t_d) \neq \sin(t-t_d) \cdot e(t-t_d) = r(t-t_d)$, so, the system is a time-varying system.

(4) 因果系统与非因果系统(causal system and non-causal system)

因果系统是指系统某时刻的输出只与系统该时刻及以前时刻的输入信号有关。例如，一个系统如果激励在 $t<t_0$ 时为零，则相应的零状态响应在 $t<t_0$ 时也应为零，这样的系统就是因果系统，否则，为非因果系统。激励可以是当前输入，也可以是历史输入或等效的初始状态。实际的物理可实现系统均为因果系统。

A system is causal if the output at any time depends only on values of the input at the present time and in the past. For causal system, if $e(t)=0$ for $t<t_0$, there must be $r(t)=0$ for $t<t_0$. The actual physical systems are causal.

Example 1-8: Determine whether or not each of the following system is causal.

① $r(t) = e(t-1) + e(1-t)$ ② $r(t) = \int_{-\infty}^{t} e(\tau) d\tau$

③ $y(n) = x(n-1) + x(n) + x(n+1)$

Solution: ① The system is not causal because the output $r(t)$ at some time may depend on future values of $e(t)$. For instance, $r(0) = e(-1) + e(1)$.

② Because the output $r(t)$ at any time depends only on values of the input at the present time and in the past. Therefore, the system is causal.

③ Because the output at any time will depend on future values of the input, so the system is also not causal.

此外，系统还可分为记忆系统与非记忆系统(systems with and without memory)，稳定系统(stable system)与不稳定系统(unstable system)等。关于系统的稳定性(stability)问题我们将在后续章节中作详细介绍。

在本书中，重点讨论线性时不变的连续时间系统[linear time invariant (LTI) continuous-time system]与线性时不变的离散时间系统[linear time invariant (LTI) discrete-time system]，它们也是系统理论的核心与基础。在本书的后续内容中，如不作特别说明的系统都是指线性时不变系统。

1.7 典型信号的 MATLAB 实现（MATLAB Realization of Typical Signals）

MATLAB 软件中提供了大量的产生基本信号的函数。最常用的指数信号、正弦信号等是 MATLAB 的内部函数，即不安装任何工具箱就可调用的函数。

1.7.1 连续信号的 MATLAB 实现(Matlab realization of continuous-time signals)

(1) 指数信号(exponential signal)

指数信号在 MATLAB 中可用 exp 函数表示，其调用形式为 y=k*exp(a*t)，图 1-30 所示因果衰减指数信号的 MATLAB 表示如下，取 $k=2$，$a=-0.5$。

k=2;
a=-0.5;
t=0:0.01:10;

y = k * exp(a * t);
plot(t, y)

(2) 抽样信号(sampling signal)

抽样信号在 MATLAB 中可用 sinc 函数表示，其调用形式为 y = sinc(t)，图 1-31 所示抽样信号的 MATLAB 表示如下：

t = -5 * pi : pi/100 : 5 * pi;
y = sinc (t/pi);
plot(t, y)

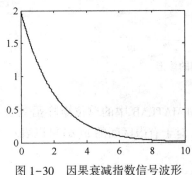

图 1-30　因果衰减指数信号波形　　　图 1-31　抽样信号波形

(3) 矩形脉冲信号(rectangular pulse signal)

矩形脉冲信号在 MATLAB 中可用 rectpuls 函数表示，其调用形式为 y = rectpuls (t, width)，用以产生一个幅度为 1，宽度为 width，以零点对称的矩形波。width 的缺省值为 1。图 1-32 所示以 $2T$ 对称的矩形脉冲信号的 MATLAB 表示如下(取 $T=1$)：

t = -0 : 0.001 : 5;
T = 1;
y = rectpuls (t-2 * T, 2 * T);
plot(t, y)

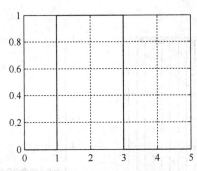

图 1-32　矩形脉冲信号的波形

1.7.2　离散信号的 MATLAB 实现(Matlab Realization of Discrete-time Signals)

(1) 指数序列(exponential sequence)

离散指数序列的一般形式为 α^n，可以用 MATLAB 中的数组幂运算 a.^k 实现。图 1-33 所示指数序列的 MATLAB 表示如下，取 $A=1$，$a=-0.5$。

n = 0 : 10;

```
A = 1; a = -0.5;
fn = A * a.^n;
stem(n, fn)
```

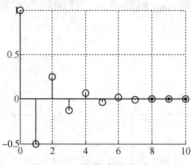

图 1-33 指数序列的波形

(2) 单位阶跃序列(uint step sequence)

单位阶跃序列 $u(n)$ 的一种简单表示方法是借助 MATLAB 中的单位矩阵函数 ones。单位矩阵 $ones(1, N)$ 产生一个由 N 个 1 组成的列向量,对于有限区间的 $u(n)$ 可以表示为

```
n = -10:10;
un = [zeros(1, 10), ones(1, 11)];
stem(n, un)
```

波形如图 1-34 所示。

(3) 单位脉冲序列(unit impulse sequence)

单位脉冲序列 $\delta(n)$ 的一种简单表示方法是借助 MATLAB 中的零矩阵函数 zeros。零矩阵 $zeros(1, N)$ 产生一个由 N 个 0 组成的列向量,对于有限区间的 $\delta(n)$ 可以表示为

```
n = -10:10;
delta = [zeros(1, 10), 1, zeros(1, 10)];
stem(n, delta)
```

波形如图 1-35 所示。

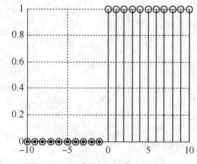

图 1-34 单位阶跃序列的波形

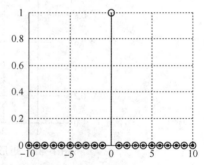

图 1-35 单位脉冲序列的波形

关键词(Key Words and Phrases)

(1) 消息   message
(2) 信息   information
(3) 信号   signal

(4) 系统　　　　　　　　　　　　system
(5) 独立变量　　　　　　　　　　independent variable
(6) 连续时间信号/系统　　　　　　continuous-time signal/system
(7) 离散时间信号/系统　　　　　　discrete-time signal/system
(8) 确定性信号　　　　　　　　　deterministic signal
(9) 随机信号　　　　　　　　　　random signal
(10) 模拟信号　　　　　　　　　 analog signal
(11) 脉冲信号　　　　　　　　　 pulse signal
(12) 数字信号　　　　　　　　　 digital signal
(13) 周期信号　　　　　　　　　 periodic signal
(14) 非周期信号　　　　　　　　 aperiodic signal
(15) 有理数　　　　　　　　　　 rational number
(16) 最小公倍数　　　　　　　　 lowest common multiple
(17) 能量信号　　　　　　　　　 energy signal
(18) 功率信号　　　　　　　　　 power signal
(19) 实指数信号　　　　　　　　 real exponential signal
(20) 复指数信号　　　　　　　　 complex exponential signal
(21) 实部　　　　　　　　　　　 real part
(22) 虚部　　　　　　　　　　　 imaginary part
(23) 直流信号　　　　　　　　　 direct current (DC) signal
(24) 正弦信号　　　　　　　　　 sinusoidal signal
(25) 抽样信号　　　　　　　　　 sampling signal
(26) 高斯信号　　　　　　　　　 Gauss signal
(27) 单位阶跃信号　　　　　　　 unit step signal
(28) 单位冲激信号　　　　　　　 unit impulse signal
(29) 单位斜坡信号　　　　　　　 unit ramp signal
(30) 理想信号　　　　　　　　　 ideal signal
(31) 符号函数(信号)　　　　　　 sign function (signal)
(32) 间断点　　　　　　　　　　 discontinuity point
(33) 序列　　　　　　　　　　　 sequence
(34) 信号变换　　　　　　　　　 signal transformation
(35) 时移　　　　　　　　　　　 time shifting
(36) 反转　　　　　　　　　　　 reflection
(37) 尺度变换　　　　　　　　　 scaling
(38) 线性/非线性系统　　　　　　linear/nonlinear system
(39) 因果/非因果系统　　　　　　causal/non-causal system
(40) 时变/时不变系统　　　　　　time-varying/time invariant system
(41) 线性时不变系统　　　　　　 linear time invariant (LTI) system
(42) 稳定/不稳定系统　　　　　　stable/unstable system
(43) 齐次性　　　　　　　　　　 scaling (homogeneity)

(44)叠加性　　　　　　　　　　　additivity
(45)微分方程　　　　　　　　　　differential equation
(46)差分方程　　　　　　　　　　difference equation
(47)双边信号　　　　　　　　　　bilateral signal
(48)单边信号　　　　　　　　　　unilateral signal

Exercises

1-1 Draw the waveform of the following signals, where, $-\infty<t<\infty$.

(1) $f(t)=u(t+1)-2u(t)+u(t-1)$ (2) $f(t)=\lim_{a\to 0}\dfrac{1}{a}[u(t)-u(t-a)]$

(3) $f(t)=\delta(t-1)-2\delta(t-2)+\delta(t-3)$ (4) $f(t)=e^{-2t}[u(t)-u(t-4)]$

(5) $f(t)=e^{-2t}\sin(2t)u(t)$ (6) $f(t)=2e^{-2t}u(t-2)$

1-2 Let $x(n)$ be a signal with $x(n)=0$ for $n<-2$ and $n>4$. For each signal given below, determine the values of n for which it is guaranteed to be zero.

(1) $x(n-3)$ 　　(2) $x(n+4)$ 　　(3) $x(-n)$ 　　(4) $x(-n+2)$

1-3 Let $f(t)$ be a signal with $f(t)=0$ for $t<3$. For each signal given below, determine the values of t for which it is guaranteed to be zero.

(1) $f(1-t)$ 　　(2) $f(1-t)+f(2-t)$ 　　(3) $f(1-t)f(2-t)$

(4) $f(3t)$ 　　(5) $f\left(\dfrac{t}{3}\right)$

1-4 Calculate the value of the integral.

(1) $\displaystyle\int_{-\infty}^{\infty}f(t-t_0)\delta(t)\,dt$ (2) $\displaystyle\int_{-\infty}^{\infty}u(t-\dfrac{t_0}{2})\delta(t-t_0)\,dt$

(3) $\displaystyle\int_{-\infty}^{t}e^{-2t}\delta(t-t_0)\,dt$ (4) $\displaystyle\int_{-\infty}^{0}e^{-2t}\delta(t-4)\,dt$

(5) $\displaystyle\int_{-\infty}^{\infty}(3t^2+t-5)\delta(2t-3)\,dt$ (6) $\displaystyle\int_{-\infty}^{\infty}e^{-j\omega t}\left[\delta\left(t+\dfrac{t_0}{2}\right)+\delta\left(t-\dfrac{t_0}{2}\right)\right]dt$

1-5 The waveform of the signal $f(t)$ is depicted in Figure 1-36, please sketch the waveform of $f(3t-4)$, $f(-3t-2)$.

Figure 1-36　the waveform of the Exercises 1-5　　　　Figure 1-37　the waveform of the Exercises 1-6

1-6 The waveform of $f(t)$ is shown in Figure 1-37, trying to sketch the waveform of the following signals.

(1) $f(3t)$ 　　(2) $f\left(\dfrac{t}{3}\right)u(3-t)$ 　　(3) $\dfrac{df(t)}{dt}$ 　　(4) $\displaystyle\int_{-\infty}^{t}f(\tau)\,d\tau$

1-7 Determine whether or not each of the following signals is periodic, if it is periodic, please determine the period.

(1) $f(t) = \sin(\pi t)$, $t \geq 0$

(2) $f(t) = \sin(2\pi t) + \cos(3\pi t + \frac{\pi}{3})$

(3) $f(t) = \sin(2t) + \cos(3\pi t)$

(4) $f(t) = e^{-2t}\sin(2t + \frac{\pi}{6})$

(5) $x(n) = \sin(\frac{3}{4}n)$

(6) $x(n) = \sin\left(\frac{\pi}{40}n + \frac{\pi}{2}\right)$

1-8 Which the following is energy signal, or power signal, or neither?

(1) $te^{-t}u(t)$

(2) $e^{t}[u(t) - u(t-1)]$

(3) $10e^{-t}\sin(t)u(t)$

(4) $\sin(t) + \sin(2\pi t)$

1-9 Determine whether or not each of the following signals is causal, linear and time invariant.

(1) $r(t) = \sin(t)e(t)u(t)$

(2) $r(t) = e'(t) + 2e(t)$

(3) $r(t) = \int_{-\infty}^{5t} e(\tau)d\tau$

(4) $y(n) = x(2n)$

(5) $r(t) = e(t-2) + e(2-t)$

(6) $r(t) = e(\frac{t}{3})$

1-10 Consider a continuous-time system with input $e(t)$ and output $r(t)$ related by
$$r(t) = e[\sin(t)]$$

(1) Is this system causal?

(2) Is this system linear?

第 2 章 连续时间系统的时域分析
(Analysis of Continuous-time System in Time Domain)

在第 1 章学习线性时不变(LTI)系统基本特性的基础上,本章主要研究 LTI 连续时间系统的时域分析法。

时域分析法(time domain analysis method)是直接研究系统时间响应或时域特性的一种方法,它以时间 t 为变量,在时域内分析系统特性。这种方法具有直观、物理概念清楚等优点,是学习各种变换分析方法的基础。

当系统采用输入、输出描述时,系统的时域解法包含两方面内容,一方面是经典法直接求解微分方程,另一方面是卷积法求解微分方程。

卷积(convolution)法是将信号分解成许多冲激信号之和,借助系统的冲激响应(impulse response),求解线性时不变系统对任意激励信号的零状态响应。对于线性时不变系统,无论是时域分析还是变换域分析,卷积运算都是重要的方法,它是联系时间域和变换域两种方法的纽带。

The convolution integral method decomposes the input signal into the sum of impulse signal, and then with the help of the impulse response of the system, the zero state response will be solved for arbitrary excitation signals. For LTI system, both the analysis of time domain and that of transform domain, the convolution operation is an important method. It is the link between time domain and transform domain.

利用经典法(classical method)求解描述系统的微分方程,这种解法将系统的全响应分为自由响应(natural response)和强迫响应(forced response)两部分,也可以按照产生响应原因的不同将系统响应分解为零输入响应(zero input response)和零状态响应(zero state response)。

By using the classical method to solve the differential equations, the complete response of the system is divided into the natural response and the forced response, also divided into the zero input response and the zero state response according to different reasons which generate the response of the system.

本章在经典法求解微分方程的基础上,重点讨论系统的零输入响应和零状态响应。通过引入冲激响应和卷积的概念,利用冲激响应和卷积求系统输出响应,使得系统分析更加简捷、明晰。

2.1 系统的时域模型(Time-domain models of system)

用数学语言描述待分析系统,建立系统的数学模型是进行系统分析的基础。本节通过对电路系统的描述,说明如何建立系统的数学模型以及 LTI 连续时间系统的一般时域模型形式。

例 2-1: 如图 2-1 所示电路,输入激励是电流源 $i_s(t)$,试列出电流 $i_L(t)$ 及 R_1 上电压 $u_1(t)$ 为输出响应变量的方程式。

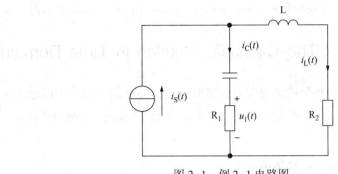

图 2-1 例 2-1 电路图

解：由 KVL，列出电压方程

$$u_C(t) + u_1(t) = u_L(t) + R_2 i_L(t) = L\frac{di_L(t)}{dt} + R_2 i_L(t) \tag{2-1}$$

对上式求导，考虑到 $i_C(t) = C\dfrac{du_C(t)}{dt}$ $R_1 i_C(t) = u_1(t)$

$$\frac{1}{R_1 C} u_1(t) + \frac{di_1(t)}{dt} = L\frac{di_L^2(t)}{dt^2} + R_2\frac{di_L(t)}{dt} \tag{2-2}$$

根据 KCL，有 $i_C(t) = i_S(t) - i_L(t)$，因而 $u_1(t) = R_1 \cdot i_C(t) = R_1 \cdot (i_S(t) - i_L(t))$

$$\frac{1}{C}(i_S(t) - i_L(t)) + R_1\left[\frac{di_S(t)}{dt} - \frac{di_L(t)}{dt}\right] = L\frac{d^2 i_L(t)}{dt^2} + R_2\frac{di_L(t)}{dt} \tag{2-3}$$

整理上式后可得

$$\frac{d^2 i_L(t)}{dt^2} + \frac{R_1 + R_2}{L} \cdot \frac{di_L(t)}{dt} + \frac{1}{LC} i_L(t) = \frac{R_1}{L}\frac{di_S(t)}{dt} + \frac{1}{LC} i_S(t) \tag{2-4}$$

$$\frac{d^2 i_1(t)}{dt^2} + \frac{R_1 + R_2}{L} \cdot \frac{di_1(t)}{dt} + \frac{1}{LC} i_1(t) = R_1\frac{d^2 i_S(t)}{dt^2} + \frac{R_1 R_2}{L} \cdot \frac{di_S(t)}{dt} \tag{2-5}$$

从上面例子可得到两点结论：

① 解得的数学模型，即求得微分方程的阶数(order)与动态电路的阶数(即独立动态元件的个数)是一致的。

② 输出响应无论是 $i_L(t)$、$u_1(t)$，或是 $u_c(t)$、$i_1(t)$，还是其他别的变量，它们的齐次方程(homogeneous equation)都相同。这表明，同一系统当它的元件参数确定不变时，它的自由频率是唯一的。

以上是代表 RC 电路系统的二阶微分方程。任何线性时不变的连续时间系统，只要给定系统结构及构成系统的各元件特性，即可写出描述该系统的输入、输出关系的线性常系数微分方程(linear constant-coefficient differential equation)式，其一般形式为

$$\sum_{i=0}^{n} a_i \frac{d^i}{dt^i} r(t) = \sum_{j=0}^{m} b_j \frac{d^j}{dt^j} e(t) \tag{2-6}$$

式中 $e(t)$ 为系统的激励信号，$r(t)$ 为系统的输出响应，n 为微分方程(系统)的阶数，$a_i(i=0, 1, \cdots, n)$、$b_j(j=0, 1, \cdots, m)$ 均为实常数。对于物理可实现系统，通常有 $n \geq m$。

Where $a_i(i=0, 1, \cdots, n)$ and $b_j(j=0, 1, \cdots, m)$ are constant coefficient of the system, $e(t)$ is the input applied to the system and $r(t)$ is the resulting output. The order of the differential equation is n, representing the number of energy storage devices in the system. Usually, $n \geq m$.

2.2 经典时域解法（The Classical Solution in Time Domain）

因为 LTI 连续时间系统的数学模型是常系数微分方程，而在高等数学中已经研究过这类方程的一般解法，它的解法由两部分构成，即齐次解（homogeneous solution）和特解（particular solution）。

2.2.1 齐次解（Homogeneous Solution）

齐次解 $r_h(t)$ 满足齐次方程（homogeneous equation）

$$\sum_{k=0}^{n} a_k \frac{d^k}{dt^k} r_h(t) = 0 \tag{2-7}$$

其形式为 $Ae^{\alpha t}$ 函数的线性组合。

将 $r_h(t) = Ae^{\alpha t}$ 代入式(2-7)得

$$\sum_{k=0}^{n} a_k \frac{d^k}{dt^k}(Ae^{\alpha t}) = \sum_{k=0}^{n} a_k \alpha^k Ae^{\alpha t} = 0$$

化简可得特征方程（characteristic equation）

$$a_n \alpha^n + a_{n-1} \alpha^{n-1} + \cdots + a_1 \alpha + a_0 = 0 \tag{2-8}$$

解此特征方程，得微分方程的特征根（characteristic roots）α_1，α_2，$\cdots$，α_n。这些特征根也称为系统的自然频率（natural frequency）或固有频率，特征根可以是 n 个不同的根，也可以是重根。

当特征根各不相同时（distinct roots），微分方程的齐次解为

$$r_h(t) = A_1 e^{\alpha_1 t} + A_2 e^{\alpha_1 t} + \cdots + A_n e^{\alpha_n t} = \sum_{i=1}^{n} A_i e^{\alpha_i t} \tag{2-9}$$

其中 A_1，A_2，$\cdots A_n$ 为由初始条件决定的系数。

当特征根有重根（multiple root）时，如 α_1 是 r 重根，则对应于 α_1 的重根部分将有 r 项，微分方程的齐次解为

$$r_h(t) = A_1 e^{\alpha_1 t} + A_2 t e^{\alpha_1 t} + \cdots + A_r t^{r-1} e^{\alpha_1 t} + \sum_{k=r+1}^{n} A_k e^{\alpha_k t} = \sum_{k=1}^{r} A_k t^{k-1} e^{\alpha_1 t} + \sum_{k=r+1}^{n} A_k e^{\alpha_k t} \tag{2-10}$$

当特征根有一对单复根（distinct complex roots）时，如 $\alpha_{1,2} = a \pm jb$，则微分方程的齐次解为

$$r_h(t) = A_1 e^{\alpha t} \cos bt + A_2 e^{\alpha t} \sin bt + \sum_{k=3}^{n} A_k e^{\alpha_k t} \tag{2-11}$$

当特征根有一对 m 重复根时，如有 m 重 $\alpha_{1,2} = a \pm jb$ 的复根，则微分方程重根对应的齐次解为

$$\begin{aligned} r_h(t) = & A_1 e^{\alpha t} \cos bt + B_1 e^{\alpha t} \sin bt + A_2 t e^{\alpha t} \cos bt + B_2 t e^{\alpha t} \sin bt + \cdots + \\ & A_m t^{m-1} e^{\alpha t} \cos bt + B_m t^{m-1} e^{\alpha t} \sin bt \end{aligned} \tag{2-12}$$

Example 2-2: Consider the differential equation for given as bellow, find the homogeneous solution.

$$\frac{d^2 r(t)}{dt^2} + 2\frac{dr(t)}{dt} + r(t) = \frac{de(t)}{dt}$$

Solution: The characteristic equation is
$$\alpha^2 + 2\alpha + 1 = 0$$
and characteristic roots is $\alpha_1 = \alpha_2 = -1$, so that the homogeneous solution is
$$r_h(t) = (A_1 + A_2 t)e^{-t}$$

2.2.2 特解(Particular Solution)

特解的函数形式与激励函数的形式有关。表 2-1 列出了常用的几种激励函数 $e(t)$ 及其所对应的特解 $r_p(t)$ 以供查用。选定特解形式后,将其代入到原微分方程,求出待定系数,即可得出特解。

The forms of the particular solution associated with common input signals are given in Table 2-1. We substitute the particular solution into differential equation, then calculate the coefficients of the function, thus obtain the particular solution.

表 2-1 几种典型激励函数对应的特解

激励函数 $e(t)$	响应函数的特解 $r_p(t)$
E(常数)	B(常数)
t^p	$b_1 t^p + b_2 t^{p-1} + \cdots + b_p t + b_{p+1}$
e^{at}	be^{at}
$\cos\omega t$	$b_1 \cos\omega t + b_2 \sin\omega t$
$\sin\omega t$	
$t^p e^{at} \cos\omega t$	$(b_1 t^p + \cdots + b_p t + b_{p+1})e^{at}\cos\omega t + (d_1 t^p + \cdots + d_p t + d_{p+1})e^{at}\sin\omega t$
$t^p e^{at} \sin\omega t$	

注:表中 B、b、d 是待定系数。

Example 2-3: Consider the system described by the following differential equation, determine the particular solution, where the input signal $e(t) = t^2$.

$$\frac{d^2 r(t)}{dt^2} + 2\frac{dr(t)}{dt} + 3r(t) = \frac{de(t)}{dt} + e(t)$$

Solution: We assume that $r_p(t) = B_2 t^2 + B_1 t + B_0$, and substitute $r_p(t)$ and $e(t) = t^2$ into the above equation, then

$$3B_2 t^2 + (4B_2 + 3B_1)t + (2B_2 + 2B_1 + 3B_0) = t^2 + 2t$$

comparing the coefficients of the equation on both sides, we can conclude

$$\begin{cases} 3B_2 = 1 \\ 4B_2 + 3B_1 = 2 \\ 2B_2 + 2B_1 + 3B_0 = 0 \end{cases}$$

that is, $B_2 = \dfrac{1}{3}$, $B_1 = \dfrac{2}{9}$, $B_0 = -\dfrac{10}{27}$

hence, the particular solution is $r_p(t) = \dfrac{1}{3}t^2 + \dfrac{2}{9}t - \dfrac{10}{27}$

2.2.3 全解(Complete Solution)

齐次解和特解相加可得系统方程的完全解。以互异特征根为例，方程的全解可写成

$$r(t) = r_h(t) + r_p(t) = A_1 e^{\alpha_1 t} + A_2 e^{\alpha_2 t} + \cdots + A_n e^{\alpha_n t} + r_p(t) \tag{2-13}$$

在系统分析中，响应区间定义为激励信号 $e(t)$ 加入后系统的状态变化区间。一般激励 $e(t)$ 都是从 $t=0$ 时刻加入，此时系统响应的求解区间为 $0^+ \leq t \leq \infty$。一组边界条件可以给定为在此区间内任一时刻 t_0，要求解满足 $r(t_0)$，$r'(t_0)$，$\cdots$，$r^{n-1}(t_0)$ 的各值。通常取 $t_0 = 0^+$，相对应的一组条件就称为初始条件(initial condition)。于是，将初始条件代入式(2-13)得

$$\begin{cases} r(0^+) = A_1 + A_2 + \cdots A_n + r_p(0^+) \\ r'(0^+) = A_1\alpha_1 + A_2\alpha_2 + \cdots A_n\alpha_n + r'_p(0^+) \\ \vdots \\ r^{n-1}(0^+) = A_1\alpha_1^{n-1} + A_2\alpha_2^{n-1} + \cdots A_n\alpha_n^{n-1} + r_p^{n-1}(0^+) \end{cases} \tag{2-14}$$

解此方程组可得齐次解中的待定系数(undetermined coefficients)。

Example 2-4: Find the complete solution of the following system. The differential equation is

$$\frac{d^2 r(t)}{dt^2} + 3\frac{dr(t)}{dt} + 2r(t) = \frac{de(t)}{dt} + 2e(t)$$

assuming that the initial values are $r(0^+) = 1$, $r'(0^+) = 1$ and the input signal $e(t) = t^2$.

Solution: The characteristic equation is

$$\alpha^2 + 3\alpha + 2 = 0 \quad \text{or} \quad (\alpha+1)(\alpha+2) = 0$$

characteristic roots are $\alpha_1 = -1$, $\alpha_2 = -2$.

① The homogeneous solution is

$$r_h(t) = A_1 e^{-t} + A_2 e^{-2t}$$

② Assume that the particular solution is

$$r_p(t) = B_2 t^2 + B_1 t + B_0$$

substitute $r_p(t)$ and $e(t)$ into the differential equation,

$$2B_2 t + (6B_2 t + 2B_1) + 2B_2 t^2 + 2B_1 t + 2B_0 = 2t^2 + 2t$$

or

$$2B_2 t^2 + (6B_2 + 2B_1)t + (2B_2 + 3B_1 + 2B_0) = 2t^2 + 2t$$

comparing the coefficients on both sides, we get

$$B_2 = 1, \quad B_1 = -2 \quad B_0 = 2$$

So that,

$$r_p(t) = t^2 - 2t + 2$$

③ The complete solution is

$$r(t) = r_h(t) + r_p(t) = A_1 e^{-t} + A_2 e^{-2t} + t^2 - 2t + 2$$

in order to determine A_1 and A_2, we substitute the initial conditions $r(0^+) = 1$, $r'(0^+) = 1$ into the complete solution, that is,

$$\begin{cases} A_1 + A_2 + 2 = 1 \\ -A_1 - 2A_2 - 2 = 1 \end{cases} \quad \text{or} \quad \begin{cases} A_1 = 1 \\ A_2 = -2 \end{cases}$$

therefore

$$r(t) = e^{-t} - 2e^{-2t} + t^2 - 2t + 2 \quad t \geq 0$$

2.2.4 从 0^- 到 0^+ 状态的转换(From 0^- State to 0^+ State)

系统在激励信号加入前瞬间的状态 $r(0^-)$, $r'(0^-)$, $r''(0^-)$, …, 称为系统的起始状态(start state), 简称 0^- 状态。起始状态包含了响应的全部过去信息, 能够反映系统中储能元件(energy storage element)的储能状况。

确定系统完全响应 $r(t)=r_h(t)+r_p(t)$ 中的待定系数是由响应区间内 $t=0^+$ 时刻的一组状态 $r(0^+)$, $r'(0^+)$, $r''(0^+)$, …确定的, 通常称这组状态为初始状态(initial state), 简称 0^+ 状态。

通常为求解描述线性时不变系统的微分方程, 就需要从已知的 0^- 状态设法求得 0^+ 状态。下面以例题说明求解方法。

From former section, we have known that 0^+ state is initial condition for solving the differential equation with classical method. The 0^+ state, however, has included the input values at $t=0$. In general, the start state of a given system often is 0^- state. How to obtain the 0^+ state from 0^- state will be discussed as follows.

在先修课程当中, 读者已经对电路系统的完全响应有所了解, 可以分为由系统起始状态引起的零输入响应(zero input response) $r_{zi}(t)$ 和由外加激励引起的零状态响应(zero state response) $r_{zs}(t)$。下面对起始值跳变的物理概念及其与数学方程的联系给出说明。

Because the complete response of a system

$$r(t) = r_{zi}(t) + r_{zs}(t) \tag{2-15}$$

therefore,

$$r(0^-) = r_{zi}(0^-) + r_{zs}(0^-) \tag{2-16}$$

$$r(0^+) = r_{zi}(0^+) + r_{zs}(0^+) \tag{2-17}$$

Because $r_{zi}(t)$ only caused by start condition, that is, by energy storage elements in the system. This means $r_{zi}(t)$ cannot jump in the interval $(0^-, 0^+)$, so $r_{zi}(0^+)$ must equal to $r_{zi}(0^-)$. Also, $r_{zs}(t)$ only caused by input signals at $t \geq 0$, so $r_{zs}(0^-)$ must be zero.

Thus,

$$r(0^+) - r(0^-) = r_{zs}(0^+) \tag{2-18}$$

or

$$r(0^+) = r(0^-) + r_{zs}(0^+) \tag{2-19}$$

$r_{zs}(0^+)$ is called jump value(跳变值)。

当系统用微分方程表示时, 从 0^- 到 0^+ 状态有没有跳变取决于微分方程右端自由项中是否包含 $\delta(t)$ 及其各阶导数项, 若包含, 则说明响应 $r(t)$ 及各阶导数发生了从 0^- 到 0^+ 状态的跳变, 即 $r(0^+) \neq r(0^-)$, $r'(0^+) \neq r'(0^-)$ 等, 这时, 如果要确定 $r(0^+)$, $r'(0^+)$ 等状态, 可以利用微分方程两端各奇异函数项的系数相平衡的方法来判断, 从而求得 0^+ 时刻的初始值。

Example 2-5: Given the following differential equation $\dfrac{dr(t)}{dt} + 3r(t) = 2\dfrac{de(t)}{dt}$, the start condition $r(0^-) = 1$, and the excitation signal $e(t) = u(t)$, find the jump value $r_{zs}(0^+)$ and initial condition $r(0^+)$.

Solution: Substitute $u(t)$ into the right side of the equation

$$\frac{dr(t)}{dt} + 3r(t) = 2\delta(t)$$

according to the coefficients balance condition of $\delta(t)$ on both sides of the equation we have

$$\frac{dr(t)}{dt} \leftrightarrow 2\delta(t)$$

or

$$r(t) \leftrightarrow 2u(t)$$

so that, $r(t)$ produced jump value from 0^- state to 0^+ state, and $r_{zs}(0^+) = 2u(0^+) = 2$.

Thus, according to the formula (2-19), $r(0^+) = r(0^-) + r_{zs}(0^+) = 1 + 2 = 3$.

Example 2-6: A system $\frac{d^2r(t)}{dt^2} + 5\frac{dr(t)}{dt} + 6r(t) = \frac{de(t)}{dt} - 2e(t)$, when $r(0^-) = 1$, $r'(0^-) = 2$, $e(t) = u(t)$, we want to know $r(0^+)$, $r'(0^+)$.

Solution: Substitute $u(t)$ into the right side of the equation

$$\frac{d^2r(t)}{dt^2} + 5\frac{dr(t)}{dt} + 6r(t) = \delta(t) - 2u(t)$$

according to the coefficients balance condition of $\delta(t)$ on both sides of the equation we can get,

$$r''(t) \leftrightarrow \delta(t)$$
$$r'(t) \leftrightarrow u(t)$$
$$r(t) \leftrightarrow t$$

so that, $r(t)$ does not jump at $t=0$ and $r'(t)$ does, that is, $r_{zs}(0^+) = 0$, $r'_{zs}(0^+) = 1$. Therefore,

$$r(0^+) = r(0^-) + r_{zs}(0^+) = 1 + 0 = 1$$
$$r'(0^+) = r'(0^-) + r'_{zs}(0^+) = 2 + 1 = 3$$

实际上,在下一节我们将看到利用 δ 函数平衡原理按经典法直接求完全解中的待定系数,可绕过从 0^- 到 0^+ 状态的过程。另外,后续课程中的拉普拉斯变换方法也可以比较简便地绕过求解 0^+ 状态的过程,直接利用 0^- 状态导出微分方程的完全解。

因此,研究 0^- 到 0^+ 状态的转换过程主要目的是从时域观察系统初始值产生跳变的物理现象,初步认识它与数学模型的对应关系,无需关注解题技巧。

2.3 LTI 连续时间系统的响应(The Response of LTI Continuous-time System)

分析信号通过系统的响应可以采用求解微分方程的经典法。但利用经典法分析系统响应存在许多局限,若描述系统的微分方程中激励项较复杂,则难以设定相应的特解形式;若激励信号发生变化,则系统响应需全部重新求解;若初始条件发生变化,则系统响应也要全部重新求解。此外,经典法是一种纯数学方法,无法突出系统响应的物理概念。在近代时域分析中,采用对系统响应进行分解的方法,能够给分析和计算带来一定的便利。

2.3.1 系统响应的分解模式(Decomposition Pattern of Complete Response)

系统的全响应(complete response)通常能够分解为自由响应与强迫响应、暂态响应与稳态响应以及零输入响应与零状态响应。

The complete response of a system usually can be divided into the natural response and forced response, transient response and steady response and zero input response and the zero state response.

（1）自由响应与强迫响应（natural response and forced response）

自由响应也称自然响应、固有响应，它反映了系统本身的特性，其形式与外加激励形式无关，取决于系统的特征根，它决定了系统自由响应的全部形式，对应于齐次解。

强迫响应也称受迫响应，只与激励函数的形式有关，利用经典法求出的特解就是系统的强迫响应。

The natural response is known as the inherent response. It reflects the characteristics of the system itself, which has nothing to do with the excitation, only depends on the characteristic roots of the system. Whereas, the forced response only depends on the excitation form.

在例 2-4 中，系统的自由响应就是 $e^{-t}-2e^{-2t}$，强迫响应就是特解 t^2-2t+2。

（2）暂态响应和稳态响应（transient response and steady response）

系统的全响应可以分解为暂态响应和稳态响应之和。当 $t\to\infty$ 时，响应等于零的那部分分量称为暂态响应或瞬态响应；当 $t\to\infty$ 时，保留下来的那部分分量称为稳态响应。暂态响应和稳态响应在系统分析中起着重要的作用。

When $t\to\infty$, the response which approaches to zero is called the transient response and the remain one is called steady response. Transient response and steady response play an important role in the analysis of system.

如果一个系统的全响应为 $r(t)=e^{-2t}+2e^{-3t}+\dfrac{1}{2}$，那么根据暂态响应和稳态响应的定义可知，系统的暂态响应为 $e^{-2t}+2e^{-3t}$，稳态响应为 $\dfrac{1}{2}$。

（3）零输入响应和零状态响应（zero input response and zero state response）

根据系统响应是仅由储能元件初始储能产生还是仅由外加激励所产生，线性时不变系统的全响应可以分解为零输入响应和零状态响应。当系统激励为零时，仅由系统的初始储能产生的响应称为系统的零输入响应，并记为 $r_{zi}(t)$。零状态响应是指系统没有初始储能，系统的起始状态为零，仅由系统的外加激励所产生的响应，记为 $r_{zs}(t)$。因此，系统的全响应为 $r(t)=r_{zi}(t)+r_{zs}(t)$。

The zero input response is only caused by the initial storage energy under the condition of without the excitation signal, denoted by $r_{zi}(t)$. The zero state response is only caused by the excitation signal but no initial storage energy, and denoted by $r_{zs}(t)$.

2.3.2 零输入响应（Zero Input Response）

若系统在 t_0 时未施加输入信号，但由于 $t<t_0$ 时系统的工作，可以使其中的储能元件蓄有能量，而能量不可能突然消失，它将逐渐释放出来，直至最后消耗殆尽。当系统的激励 $e(t)$ 为零，仅由系统的初始储能产生的响应称为系统的零输入响应，并记为 $r_{zi}(t)$。

零输入响应 $r_{zi}(t)$ 是满足方程 $a_n\dfrac{d^n r(t)}{dt^n}+a_{n-1}\dfrac{d^{n-1}r(t)}{dt^{n-1}}+\cdots+a_1\dfrac{dr(t)}{dt}+a_0 r(t)=0$ 及起始状态 $r^{(k)}(0^-)$（$k=0,1,\cdots,n-1$）的解，对应齐次微分方程的齐次解。

Example 2-7: The system is described by the differential equation

$$\frac{d^2r(t)}{dt^2} + 2\frac{dr(t)}{dt} + 5r(t) = e(t)$$

find the zero input response, assume that $r(0^-) = 1$, $r'(0^-) = 7$.

Solution: Because the zero input response $r_{zi}(t)$ is a solution of the homogeneous differential equation $\frac{d^2r(t)}{dt^2} + 2\frac{dr(t)}{dt} + 5r(t) = 0$, and the characteristic equation is

$$\alpha^2 + 2\alpha + 5 = 0$$

so the characteristic roots are $\alpha_1 = -1 + 2j$, $\alpha_2 = -1 - 2j$

therefore, the form of the zero input response $r_{zi}(t)$ is

$$r_{zi}(t) = e^{-t}(c_1\cos 2t + c_2\sin 2t)$$

substituting the initial conditions $r(0^-) = 1$, $r'(0^-) = 7$ into the above equation, we get

$$r_{zi}(0^-) = c_1 = 1$$
$$r'_{zi}(t) = -e^{-t}[c_1\cos 2t + c_2\sin 2t] + e^{-t}[-2c_1\sin 2t + 2c_2\cos 2t]$$
$$r'_{zi}(0^-) = -c_1 + 2c_2 = 7 \Rightarrow c_2 = 4$$

So that the zero input response is

$$r_{zi}(t) = e^{-t}[\cos 2t + 4\sin 2t] \quad t \geq 0$$

虽然零输入响应和自由响应都是齐次方程的解,但二者的系数各不相同。零输入响应的系数仅由系统的起始状态决定,而自由响应的系数要由系统的起始状态和激励共同来确定。

2.3.3 零状态响应(Zero State Response)

所谓零状态是指系统没有初始储能,系统的起始状态为零,即

$$r(0^-) = r^{(1)}(0^-) = \cdots = r^{(n-1)}(0^-) = 0$$

这时仅由系统的外加激励所产生的响应称为零状态响应,并记为 $r_{zs}(t)$。

零状态响应 $r_{zs}(t)$ 由起始状态为零时的方程

$$\begin{cases} a_n\dfrac{d^nr(t)}{dt^n} + a_{n-1}\dfrac{d^{n-1}r(t)}{dt^{n-1}} + \cdots + a_1\dfrac{dr(t)}{dt} + a_0r(t) = \\ b_m\dfrac{d^me(t)}{dt^m} + b_{m-1}\dfrac{d^{m-1}e(t)}{dt^{m-1}} + \cdots + b_1\dfrac{de(t)}{dt} + b_0e(t) \\ r^{(k)}(0^-) = 0, \quad k = 0, 1, \cdots, n-1 \end{cases} \quad (2\text{-}20)$$

所确定。

零状态响应的经典解法将其分解为

$$r_{zs}(t) = r_{zsh}(t) + r_{zsp}(t) \quad (2\text{-}21)$$

其中 $r_{zsh}(t)$ 和 $r_{zsp}(t)$ 分别为式(2-20)的齐次解和特解。

Example 2-8: Consider the differential equation

$$\frac{dr(t)}{dt} + r(t) = e(t) \quad (2\text{-}22)$$

determine the zero state response $r_{zs}(t)$, assuming that $e(t) = \cos t\, u(t)$ and the initial condition is at rest.

Solution: Let $r_{zs}(t) = r_{zsh}(t) + r_{zsp}(t)$,

where,
$$r_{zsh}(t) = Ae^{-t}, \text{ and } r_{zsp}(t) = B_1\cos t + B_2\sin t$$

so that the zero state response is
$$r_{zs}(t) = Ae^{-t} + B_1\cos t + B_2\sin t$$

because of the condition of initial rest, $r(t) = 0$ for $t<0$, we rewrite the form of $r_{zs}(t)$
$$r_{zs}(t) = (Ae^{-t} + B_1\cos t + B_2\sin t)u(t)$$

substituting this into Eq. (2-22), we obtain
$$(A+B_1)\delta(t) + [-Ae^{-t} - B_1\sin t + B_2\cos t]u(t) + [Ae^{-t} + B_1\cos t + B_2\sin t]u(t) = \cos t\, u(t)$$

or
$$(A+B_1)\delta(t) + [(B_1+B_2)\cos t + (B_2-B_1)\sin t]u(t) = \cos t\, u(t)$$

therefore,
$$\begin{cases} B_1 + B_2 = 1 \\ B_2 - B_1 = 0 \\ A + B_1 = 0 \end{cases}$$

then $B_1 = \dfrac{1}{2}$, $B_2 = \dfrac{1}{2}$, $A = -\dfrac{1}{2}$

hence, the zero state response
$$r_{zs}(t) = \left(-\frac{1}{2}e^{-t} + \frac{1}{2}\cos t + \frac{1}{2}\sin t\right)u(t)$$

将系统的响应分解为零输入响应和零状态响应时，求解零输入响应相对比较简单，只需在方程齐次解的基础上，利用系统的起始状态得出齐次解的待定系数即可，但求系统的零状态响应并不容易，此时，不仅需要求出方程的齐次解和特解，而且需要确定解中的待定系数。

求解系统的零状态响应一种比较简便的方法是利用激励与系统冲激响应的卷积(convolution)来实现，在本章2.4节中我们将做详细介绍。

2.3.4 冲激响应(Unit Impulse Response)

连续系统的冲激响应定义为在系统起始状态为零的条件下，以单位冲激信号激励系统所产生的输出响应，以符号$h(t)$表示。由于系统冲激响应$h(t)$要求系统在零状态条件下，且输入激励为单位冲激信号$\delta(t)$，因而冲激响应$h(t)$仅取决于系统的内部结构及其元件参数。因此，系统的冲激响应$h(t)$可以表征系统本身的特性。换句话说，不同的系统就会有不同的冲激响应。连续时间LTI系统的冲激响应在求解系统零状态响应$r_{zs}(t)$中起着十分重要的作用。因此，冲激响应$h(t)$的分析是系统分析的重要内容。

The impulse response $h(t)$ is the output of an LTI system when $\delta(t)$ is the input. The zero state response of an LTI system can be deduced by the impulse response of the system. Therefore, we must discuss the methods to find the impulse response from differential equation of the system.

根据连续时间LTI系统的数学模型，其冲激响应$h(t)$满足微分方程
$$\sum_{i=0}^{n} a_i \frac{d^i}{dt^i}h(t) = \sum_{j=0}^{m} b_j \frac{d^j}{dt^j}\delta(t) \tag{2-23}$$

及初始状态$h^{(i)}(0^-) = 0$ ($i = 0, 1, \cdots n-1$)。由于$\delta(t)$及其各阶导数在$t \geq 0^+$时都等于零，故式(2-23)右端各项在$t \geq 0^+$时恒等于零，这时式(2-23)成为齐次方程，这样冲激响应$h(t)$

的形式应与齐次解的形式相同(由特征根的形式来决定,分为互异根、重根、共轭复根等几种情况)。

如系统的特征根是互异实根,且当 n>m 时,$h(t)$ 可以表示为

$$h(t) = \sum_{i=1}^{n} k_i e^{\alpha_i t} u(t) \qquad (2\text{-}24)$$

式中的待定系数 $k_i(i=1, 2, \cdots, n)$ 可以采用冲激平衡法(impulse balance method)确定,即将式(2-24)代入式(2-23)中,为保持系统对应的微分方程式恒等,方程式两边所具有的冲激信号及其高阶导数必须相等,根据此规则即可求出系统的冲激响应 $h(t)$ 中的待定系数。当 n≤m 时,要使方程式两边所具有的冲激信号及其高阶导数(higher order derivative)相等,则 $h(t)$ 表达式中还应含有 $\delta(t)$ 及其相应阶的导数 $\delta^{(m-n)}(t)$,$\delta^{(m-n-1)}(t)$,$\cdots$,$\delta'(t)$ 等项。下面举例说明冲激响应的求解。

Example 2-9: Consider the differential equation

$$\frac{dr(t)}{dt} + 4r(t) = 2e(t)$$

try to find the impulse response.

Solution: According to the definition of the impulse response, when $e(t)=\delta(t)$, then $h(t)$ is satisfied to the following equation

$$\frac{dh(t)}{dt} + 4h(t) = 2\delta(t) \qquad (2\text{-}25)$$

because the characteristic root $\alpha=-4$ and n>m, so

$$h(t) = k e^{-4t} u(t)$$

substituting $h(t) = k e^{-4t} u(t)$ into Eq. (2-25), we have

$$k e^{-4t} \delta(t) - 4 k e^{-4t} u(t) + 4 k e^{-4t} u(t) = 2\delta(t)$$

that is, $\qquad k e^{-4t} \delta(t) = k \delta(t) = 2\delta(t) \Rightarrow k = 2$

Thus, $\qquad h(t) = 2 e^{-4t} u(t)$

注意,在对 $k e^{\alpha t} u(t)$ 进行求导时,必须按两个函数乘积的导数公式进行。求导后,对含有 $\delta(t)$ 的项利用冲激信号的筛选特性进行化简,即

$$f(t)\delta(t) = f(0)\delta(t)$$

Example 2-10: Consider the differential equation

$$\frac{dr(t)}{dt} + 5r(t) = 2\frac{de(t)}{dt}$$

determine the impulse response of the system.

Solution: For this differential equation, where, n=m, thus,

$$h(t) = k_1 e^{-5t} u(t) + k_2 \delta(t)$$

substituting $h(t)$ and $e(t) = \delta(t)$ into the differential equation, we get

$$-5 k_1 e^{-5t} u(t) + k_1 \delta(t) + k_2 \delta'(t) + 5 k_1 e^{-5t} u(t) + 5 k_2 \delta(t) = 2\delta'(t)$$

or

$$k_2 \delta'(t) + (k_1 + 5 k_2)\delta(t) = 2\delta'(t)$$

comparing the coefficients on both sides, we have $k_2 = 2$, $k_1 = -10$, thus,

$$h(t) = -10 e^{-5t} u(t) + 2\delta(t)$$

2.3.5 阶跃响应(Unit Step Response)

对于 LTI 系统，当其初始状态为零时，输入为单位阶跃信号所引起的响应称为单位阶跃响应，简称阶跃响应，用 $g(t)$ 表示。阶跃响应是激励为单位阶跃信号 $u(t)$ 时，系统的零状态响应。

The step response of an LTI system is the response if the input applied to the system is the unit step. For the linear time invariant system, because of the relationship between the unit impulse signal and the unit step signal, the same relationship exists between the impulse response and the step response, that is,

$$\begin{cases} h(t) = \dfrac{\mathrm{d}}{\mathrm{d}t} g(t) \\ g(t) = \int_{-\infty}^{t} h(\tau)\mathrm{d}\tau \end{cases} \tag{2-26}$$

时域法求系统的冲激响应 $h(t)$ 与阶跃响应 $g(t)$ 方法比较直观，物理概念明确，是变换域分析的基础。在学习拉普拉斯变换后，用变换域分析方法求 $h(t)$ 和 $g(t)$ 将更为简洁方便。

2.4 卷积积分(The Convolution Integral)

随着信号与系统理论研究的深入及计算机技术的发展，卷积运算越来越受到重视。在现代地震勘探、超声诊断、光学成像、系统辨识及其他诸多信号处理领域中，卷积运算得到了广泛的应用，而且许多都是有待深入研究的课题。

本节将先对卷积积分的概念、运算方法加以说明，然后阐述卷积的基本性质以及如何利用卷积求解系统的零状态响应。

The convolution integral plays an important role in system analysis. In this section, we will introduce its concept, operation and several useful properties, and then the method of solving the zero state response using the convolution integral will also be discussed.

2.4.1 卷积积分的概念(Concept of the Convolution Integral)

一般情况下，如有任意两个信号 $f_1(t)$ 和 $f_2(t)$ 做运算

$$f(t) = \int_{-\infty}^{\infty} f_1(\tau) f_2(t-\tau)\mathrm{d}\tau \tag{2-27}$$

则此运算定义为 $f_1(t)$ 和 $f_2(t)$ 的卷积(convolution)，简记为

$$f(t) = f_1(t) * f_2(t) \tag{2-28}$$

这里对 $f_1(t)$ 和 $f_2(t)$ 的作用时间没有加以限制，积分上下限取 $-\infty$ 和 ∞。在实际工程中，由于系统的因果性或激励存在时间的局限性，积分上下限会有所变化，在运算中一定要加以注意。

Eq. (2-27) is referred to as the convolution integral or the superposition integral of two continuous-time signals $f_1(t)$ and $f_2(t)$. For convenience, the convolution of two signals $f_1(t)$ and $f_2(t)$ will be represented symbolically as $f(t) = f_1(t) * f_2(t)$. The symbol $*$ denote both discrete-time and continuous-time convolution.

2.4.2 卷积积分的图解法(Graphic Method of Convolution Integral)

如果已知信号$f_1(t)$和$f_2(t)$的波形，用图解法能直观地说明卷积积分的计算过程，而且便于理解卷积积分概念。两个信号$f_1(t)$和$f_2(t)$的卷积运算可通过以下几个步骤来完成：

第一步：换元，将波形图中的t轴改换成τ轴，分别得到$f_1(\tau)$和$f_2(\tau)$；

第二步：反转，将$f_2(\tau)$波形以纵轴为中心翻转180°，得到$f_2(-\tau)$；

第三步：平移，给定一个t值，将$f_2(-\tau)$波形沿τ轴平移t。当$t<0$时，波形往左移，当$t>0$时，波形往右移，得到$f_2(t-\tau)$；

第四步：乘积，将$f_1(\tau)$和$f_2(t-\tau)$相乘，得到卷积积分式中的被积函数$f_1(\tau)f_2(t-\tau)$；

第五步：积分，令变量t在$(-\infty,+\infty)$范围内变化，计算积分$f(t)=\int_{-\infty}^{\infty}f_1(\tau)f_2(t-\tau)\mathrm{d}\tau$。

Example 2-11: Consider the convolution of the following two signals, which are depicted in Figure 2-2(a).

Solution: From the definition of the convolution integral of two continuous-time signals,

$$f(t) = \int_{-\infty}^{\infty} f_1(\tau)f_2(t-\tau)\mathrm{d}\tau$$

① For $t<0$, there is no overlap between the nonzero portions of $f_1(\tau)$ and $f_2(t-\tau)$, and consequently, $f(t)=0$. The signal $f_1(\tau)$ and $f_2(t-\tau)$ are plotted as functions of τ in Figure 2-2(b).

② In Figure 2-2(c), for $0 \leq t < T$, $f_1(\tau)f_2(t-\tau) = \begin{cases} t-\tau, & 0 \leq \tau \leq t \\ 0, & \text{otherwise} \end{cases}$, thus, in this interval,

$$f(t) = \int_0^t (t-\tau)\mathrm{d}\tau = \frac{1}{2}t^2$$

③ For $t \geq T$, but $t-2T<0$, i.e. $T \leq t < 2T$, see Figure 2-2(d),

$$f_1(\tau)f_2(t-\tau) = \begin{cases} t-\tau, & 0 \leq \tau \leq T \\ 0, & \text{otherwise} \end{cases}$$

thus, in this interval,

$$f(t) = \int_0^T (t-\tau)\mathrm{d}\tau = Tt - \frac{1}{2}T^2$$

④ As shown in Figure 2-2(e), for $t-2T \geq 0$, but $t-2T<T$, i.e. $2T \leq t < 3T$,

$$f_1(\tau)f_2(t-\tau) = \begin{cases} t-\tau, & t-2T \leq \tau \leq T \\ 0, & \text{otherwise} \end{cases}$$

thus, in this interval,

$$f(t) = \int_{t-2T}^T (t-\tau)\mathrm{d}\tau = -\frac{1}{2}t^2 + Tt + \frac{3}{2}T^2$$

⑤ As Figure 2-2(f) shows, for $t-2T \geq T$, or equivalently, $t \geq 3T$, there is no overlap between the nonzero portions of $f_1(\tau)$ and $f_2(t-\tau)$, and hence, $f(t)=0$.

Summarizing, the convolution of $f_1(t)$ and $f_2(t)$ is

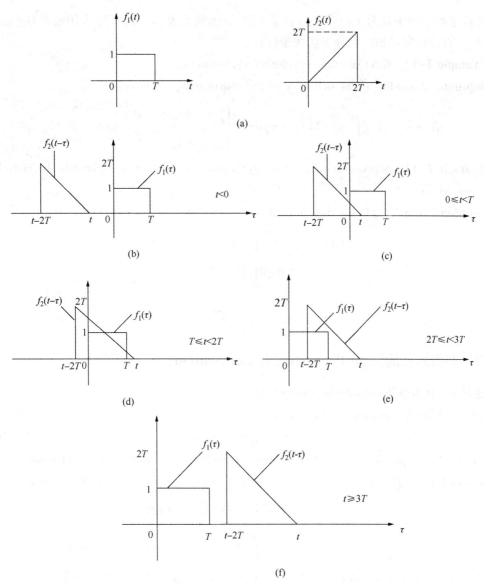

Figure 2-2　product $f_1(\tau)f_2(t-\tau)$ for different values of t for Example 2-11

$$f(t) = \begin{cases} 0, & t < 0 \\ \dfrac{1}{2}t^2, & 0 \leq t < T \\ Tt - \dfrac{1}{2}T^2, & T \leq t < 2T \\ -\dfrac{1}{2}t^2 + Tt + \dfrac{3}{2}T^2, & 2T \leq t < 3T \\ 0, & 3T \leq t \end{cases}$$

从卷积的计算过程可以清楚地看到，确定积分区间与被积函数表达式是其关键。卷积结果 $f(t)$ 的起点等于 $f_1(t)$ 和 $f_2(t)$ 的起点之和，$f(t)$ 的终点等于 $f_1(t)$ 和 $f_2(t)$ 的终点之和。若卷积的两个信号不含有冲激信号或其各阶导数，则卷积的结果必定为一个连续函数，不会出现间断点。此外，翻转信号时，尽可能翻转比较简单的信号，以简化运算过程。

若待卷积的两个信号 $f_1(t)$ 和 $f_2(t)$ 能用解析函数式表达，则可以采用解析法（analytic method），直接按照卷积的积分表达式来计算。

Example 2-12：Calculate the convolution $u(t) * u(t)$.

Solution：According to the definition of the convolution, we have

$$u(t)*u(t) = \int_{-\infty}^{\infty} u(\tau)u(t-\tau)d\tau = \begin{cases} \int_0^t 1 \cdot d\tau, & t>0 \\ 0, & t \leq 0 \end{cases} = tu(t) = R(t)$$

Example 2-13：Suppose that $f_1(t) = e^{-3t}u(t)$ and $f_2(t) = e^{-5t}u(t)$, calculate the convolution of the two signals.

Solution：According to the definition of the convolution, we have

$$f_1(t)*f_2(t) = \int_{-\infty}^{\infty} f_1(\tau)f_2(t-\tau)d\tau$$

$$= \int_{-\infty}^{\infty} e^{-3\tau}u(\tau) \cdot e^{-5(t-\tau)}u(t-\tau)d\tau$$

$$= e^{-5t}\int_0^t e^{2\tau}d\tau \cdot u(t) = \frac{1}{2}(e^{-3t} - e^{-5t})u(t)$$

2.4.3 卷积运算的性质（Properties of Convolution）

性质1 代数性质（algebraic properties）

（1）交换律（the commutative property）

$$f_1(t)*f_2(t) = f_2(t)*f_1(t) \tag{2-29}$$

Proof：This expression can be verified in a straightforward manner by means of a substitution of variables in Eq. (2-27). For example, if we let $t-\tau=x$ or, equivalently, $\tau=t-x$, then

$$f_1(t)*f_2(t) = \int_{-\infty}^{\infty} f_1(\tau)f_2(t-\tau)d\tau$$

$$= \int_{-\infty}^{\infty} f_1(t-x)f_2(x)dx = f_2(t)*f_1(t)$$

在 $f_1(t)*f_2(t)$ 进行积分计算繁琐的情况下，采用交换律有时会使运算简便准确。应用在工程上，卷积运算的交换律还可说明一个单位冲激响应是 $h(t)$ 的 LTI 系统对输入信号 $e(t)$ 所产生的响应，与一个单位冲激响应是 $e(t)$ 的 LTI 系统对输入信号 $h(t)$ 所产生的响应相同。

（2）结合律（the associative property）

$$[f_1(t)*f_2(t)]*f_3(t) = f_1(t)*[f_2(t)*f_3(t)] \tag{2-30}$$

Proof：this property is proven by straightforward manipulation of the integrals involved.

$$[f_1(t)*f_2(t)]*f_3(t) = \int_{-\infty}^{\infty} f_1(\tau)f_2(t-\tau)d\tau * f_3(t)$$

$$= \int_{-\infty}^{\infty} f_1(\tau)[f_3(t)*f_2(t-\tau)]d\tau$$

$$= \int_{-\infty}^{\infty} f_1(\tau)\left[\int_{-\infty}^{\infty} f_3(\lambda)f_2(t-\lambda-\tau)d\lambda\right]d\tau$$

$$= \int_{-\infty}^{\infty} f_1(\tau)\left[\int_{-\infty}^{\infty} f_3(\lambda)f_2((t-\tau)-\lambda)d\lambda\right]d\tau$$

let $\int_{-\infty}^{\infty} f_3(\lambda) f_2((t-\tau)-\lambda) d\lambda = g(t-\tau)$, so that

$$[f_1(t) * f_2(t)] * f_3(t) = \int_{-\infty}^{\infty} f_1(\tau) \left[\int_{-\infty}^{\infty} f_3(\lambda) f_2((t-\tau)-\lambda) d\lambda \right] d\tau$$

$$= \int_{-\infty}^{\infty} f_1(\tau) \cdot g(t-\tau) d\tau$$

$$= f_1(t) * g(t)$$

where, $g(t) = \int_{-\infty}^{\infty} f_3(\lambda) f_2(t-\lambda) d\lambda = f_3(t) * f_2(t)$

(3) 分配律(the distributive property)

$$f_1(t) * [f_2(t) + f_3(t)] = f_1(t) * f_2(t) + f_1(t) * f_3(t) \tag{2-31}$$

Proof: This property can be verified in a straightforward manner.

$$f_1(t) * [f_2(t) + f_3(t)] = \int_{-\infty}^{\infty} f_1(\tau) [f_2(t-\tau) + f_3(t-\tau)] d\tau$$

$$= \int_{-\infty}^{\infty} f_1(\tau) f_2(t-\tau) d\tau + \int_{-\infty}^{\infty} f_1(\tau) f_3(t-\tau) d\tau$$

$$= f_1(t) * f_2(t) + f_1(t) * f_3(t)$$

性质2 任意信号与奇异信号的卷积(the convolution of arbitrary signal and singular signal)

(1) 任意信号与冲激信号的卷积(the convolution of arbitrary signal and impulse signal)

$$f(t) * \delta(t) = f(t) \tag{2-32}$$

$$f(t) * \delta(t-t_1) = f(t-t_1) \tag{2-33}$$

$$f(t) * \delta'(t) = f'(t) \tag{2-34}$$

Proof: According to the definition of the convolution and the sampling property of $\delta(t)$, we can deduce

$$f(t) * \delta(t) = \delta(t) * f(t) = \int_{-\infty}^{\infty} \delta(\tau) f(t-\tau) d\tau$$

$$= f(t) \int_{-\infty}^{\infty} \delta(\tau) d\tau = f(t)$$

similarly, the convolution of $f(t)$ and the shifting of $\delta(t)$ and the impulse coupling can be verified simply.

Example 2-14: Suppose $f(t) = f_1(t) * f_2(t)$, prove the equality

$$f_1(t-t_1) * f_2(t-t_2) = f(t-t_1-t_2)$$

Proof: $f_1(t-t_1) * f_2(t-t_2) = [f_1(t) * \delta(t-t_1)] * [f_2(t) * \delta(t-t_2)]$

$$= f_1(t) * \delta(t-t_1) * f_2(t) * \delta(t-t_2)$$

$$= f_1(t) * f_2(t) * [\delta(t-t_1) * \delta(t-t_2)]$$

$$= f(t) * \delta(t-t_1-t_2)$$

$$= f(t-t_1-t_2)$$

(2) 任意信号与阶跃信号的卷积(the convolution of arbitrary signal and step signal)

$$f(t) * u(t) = \int_{-\infty}^{t} f(\tau) d\tau \tag{2-35}$$

Proof: According to the definition of the convolution

$$f(t) * u(t) = \int_{-\infty}^{\infty} f(\tau) u(t - \tau) \mathrm{d}\tau = \int_{-\infty}^{t} f(\tau) \mathrm{d}\tau$$

Example 2-15: Calculate the convolution $u(t) * u(t)$.

Solution: According to Eq. (2-35), we can get $u(t) * u(t) = \int_{-\infty}^{t} u(\tau) \mathrm{d}\tau = R(t)$.

性质 3 卷积的微分与积分 (differential and integral of the convolution)

(1) 微分性质 (differential property)

$$\frac{\mathrm{d}}{\mathrm{d}t}[f_1(t) * f_2(t)] = \frac{\mathrm{d}}{\mathrm{d}t} f_1(t) * f_2(t) = f_1(t) * \frac{\mathrm{d}}{\mathrm{d}t} f_2(t) \quad (2-36)$$

Proof: This property can be verified in a straightforward manner.

$$\frac{\mathrm{d}}{\mathrm{d}t}[f_1(t) * f_2(t)] = \frac{\mathrm{d}}{\mathrm{d}t}\left[\int_{-\infty}^{\infty} f_1(\tau) f_2(t - \tau) \mathrm{d}\tau\right]$$

$$= \int_{-\infty}^{\infty} f_1(\tau) \frac{\mathrm{d}}{\mathrm{d}t}[f_2(t - \tau)] \mathrm{d}\tau = f_1(t) * \frac{\mathrm{d}}{\mathrm{d}t} f_2(t)$$

similarly, $\frac{\mathrm{d}}{\mathrm{d}t}[f_1(t) * f_2(t)] = \frac{\mathrm{d}}{\mathrm{d}t} f_1(t) * f_2(t)$.

(2) 积分性质 (integral property)

$$\int_{-\infty}^{t} f_1(\lambda) * f_2(\lambda) \mathrm{d}\lambda = f_1(t) * \int_{-\infty}^{t} f_2(\lambda) \mathrm{d}\lambda = \int_{-\infty}^{t} f_1(\lambda) \mathrm{d}\lambda * f_2(t) \quad (2-37)$$

Proof: According to the definition of the convolution

$$\int_{-\infty}^{t} f_1(\lambda) * f_2(\lambda) \mathrm{d}\lambda = \int_{-\infty}^{t} \left[\int_{-\infty}^{\infty} f_1(\tau) f_2(\lambda - \tau) \mathrm{d}\tau\right] \mathrm{d}\lambda$$

$$= \int_{-\infty}^{\infty} f_1(\tau) \left[\int_{-\infty}^{t} f_2(\lambda - \tau) \mathrm{d}\lambda\right] \mathrm{d}\tau$$

$$= \int_{-\infty}^{\infty} f_1(\tau) \left[\int_{-\infty}^{t-\tau} f_2(x) \mathrm{d}x\right] \mathrm{d}\tau$$

$$= f_1(t) * \int_{-\infty}^{t} f_2(\lambda) \mathrm{d}\lambda$$

similarly, $\int_{-\infty}^{t} f_1(\lambda) * f_2(\lambda) \mathrm{d}\lambda = \int_{-\infty}^{t} f_1(\lambda) \mathrm{d}\lambda * f_2(t)$.

(3) 微分-积分性质 (differential-integral property)

$$\frac{\mathrm{d}}{\mathrm{d}t} f_1(t) * \int_{-\infty}^{t} f_2(\lambda) \mathrm{d}\lambda = f_1(t) * f_2(t) \quad (2-38)$$

Combining the differential property and integral property, this result is easy to verified as follow

$$\frac{\mathrm{d}}{\mathrm{d}t} f_1(t) * \int_{-\infty}^{t} f_2(\lambda) \mathrm{d}\lambda = \frac{\mathrm{d}}{\mathrm{d}t}\left[f_1(t) * \int_{-\infty}^{t} f_2(\lambda) \mathrm{d}\lambda\right]$$

$$= f_1(t) * \frac{\mathrm{d}}{\mathrm{d}t}\left[\int_{-\infty}^{t} f_2(\lambda) \mathrm{d}\lambda\right]$$

$$= f_1(t) * f_2(t)$$

Example 2-16: Using the differential-integral property to calculate the convolution $u(t) * u(t)$.

Solution: From Eq. (2-38),

$$u(t) * u(t) = \frac{\mathrm{d}}{\mathrm{d}t}u(t) * \int_{-\infty}^{t} u(\tau)\mathrm{d}\tau = \delta(t) * R(t) = R(t)$$

注意：应用上述公式时 $f_1(t)$ 和 $f_2(t)$ 必须是可微、可积函数。

2.4.4 用卷积法求解系统的零状态响应(Solving the Zero State Response Using the Convolution)

(1) 信号分解为冲激信号的线性组合(the representation of signals in terms of impulses)

在信号分析与系统分析时，常常需要将信号分解为基本信号的形式。这样，对信号与系统的分析就变为对基本信号的分析，从而将复杂问题简单化，且可以使信号与系统分析的物理过程更加清晰。信号分解为冲激信号序列就是其中的一个实例，如图 2-3 所示。

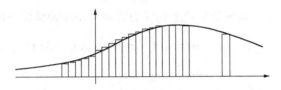

图 2-3 任意信号分解为冲激信号序列

从上图可见，将任意信号 $f(t)$ 分解成许多小矩形，间隔为 $\Delta\tau$，各矩形的高度就是信号 $f(t)$ 在该点的函数值。根据函数积分原理，当 $\Delta\tau$ 很小时，可以用这些小矩形的顶端构成阶梯信号来近似表示信号 $f(t)$；而当 $\Delta\tau \to 0$ 时，可以用这些小矩形来精确表达信号 $f(t)$。即

$$f(t) \approx \cdots + f(0)(u(t) - u(t-\Delta\tau)) + f(\Delta\tau)(u(t-\Delta\tau) - u(t-2\Delta\tau)) + \cdots + f(k\Delta\tau)(u(t-k\Delta\tau) - u(t-k\Delta\tau - \Delta\tau)) + \cdots$$

$$= \cdots + f(0)\frac{u(t) - u(t-\Delta\tau)}{\Delta\tau}\Delta\tau + f(\Delta\tau)\frac{u(t-\Delta\tau) - u(t-2\Delta\tau)}{\Delta\tau}\Delta\tau + \cdots + f(k\Delta\tau)\frac{u(t-k\Delta\tau) - u(t-k\Delta\tau - \Delta\tau)}{\Delta\tau}\Delta\tau + \cdots$$

$$= \sum_{k=-\infty}^{\infty} f(k\Delta\tau)\frac{u(t-k\Delta\tau) - u(t-k\Delta\tau - \Delta\tau)}{\Delta\tau}\Delta\tau$$

上式只是近似表示信号 $f(t)$，且 $\Delta\tau$ 越小，其误差越小。当 $\Delta\tau \to 0$ 时，可以用上式精确地表示信号 $f(t)$。由于当 $\Delta\tau \to 0$ 时，$k\Delta\tau \to \tau$，$\Delta\tau \to \mathrm{d}\tau$，且

$$\frac{u(t-k\Delta\tau) - u(t-k\Delta\tau - \Delta\tau)}{\Delta\tau} \to \delta(t-\tau)$$

故在 $\Delta\tau \to 0$ 时，

$$f(t) = \lim_{\Delta\tau \to 0}\sum_{k=-\infty}^{\infty} f(k\Delta\tau)\frac{u(t-k\Delta\tau) - u(t-k\Delta\tau - \Delta\tau)}{\Delta\tau} \cdot \Delta\tau$$

$$= \lim_{\Delta\tau \to 0}\sum_{k=-\infty}^{\infty} f(k\Delta\tau) \cdot \delta(t-k\Delta\tau) \cdot \Delta\tau$$

(2-39)

(2) 卷积法求零状态响应(solving the zero state response using the convolution)

设施加于线性时不变系统的任意激励是 $e(t)$，该系统的冲激响应为 $h(t)$，由式(2-39)可知：任意信号 $e(t)$ 可以分解为无限多个冲激信号的叠加。不同的信号 $e(t)$ 只是冲激信号 $\delta(t-k\Delta\tau)$ 前的系数 $e(k\Delta\tau)$ 不同(系数亦即是该冲激信号的强度)。这样，任一信号 $e(t)$ 作用

于系统产生的响应 $r_{zs}(t)$ 可由各个 $\delta(t-k\Delta\tau)$ 产生的响应叠加而成，即有下列关系式成立：
$$\delta(t) \Rightarrow h(t)$$
表示冲激信号 $\delta(t)$ 作用于 LTI 系统所产生的响应为 $h(t)$，以下类似。
$$\delta(t-k\Delta\tau) \Rightarrow h(t-k\Delta\tau)$$
$$e(k\Delta\tau)\Delta\tau \cdot \delta(t-k\Delta\tau) \Rightarrow e(k\Delta\tau)\Delta\tau \cdot h(t-k\Delta\tau)$$
$$\sum_{k=-\infty}^{\infty} e(k\Delta\tau)\Delta\tau \cdot \delta(t-k\Delta\tau) \Rightarrow \sum_{k=-\infty}^{\infty} e(k\Delta\tau)\Delta\tau \cdot h(t-k\Delta\tau)$$
$$\lim_{\Delta\tau \to 0}\sum_{k=-\infty}^{\infty} e(k\Delta\tau)\delta(t-k\Delta\tau) \cdot \Delta\tau \Rightarrow \lim_{\Delta\tau \to 0}\sum_{k=-\infty}^{\infty} e(k\Delta\tau)h(t-k\Delta\tau) \cdot \Delta\tau = \int_{-\infty}^{\infty} e(\tau)h(t-\tau)\mathrm{d}\tau$$
$$e(t) \Rightarrow r_{zs}(t) = \int_{-\infty}^{\infty} e(\tau)h(t-\tau)\mathrm{d}\tau$$

因此，对于任一时刻，系统的零状态响应为激励与系统冲激响应的卷积，即
$$r_{zs}(t) = \int_{-\infty}^{\infty} e(\tau)h(t-\tau)\mathrm{d}\tau = e(t) * h(t) \tag{2-40}$$

Examples 2-17: Consider the following system,
$$\frac{\mathrm{d}^2 r(t)}{\mathrm{d}t^2} + 3\frac{\mathrm{d}r(t)}{\mathrm{d}t} + 2r(t) = \frac{\mathrm{d}e(t)}{\mathrm{d}t} + e(t)$$
if the input $e(t)=(2e^{-t}-1)u(t)$ and the initial conditions are $r(0^-)=r'(0^-)=1$, determine $r_{zi}(t)$, $h(t)$, $r_{zs}(t)$ and the complete response $r(t)$.

Solution: ① Find the zero input response $r_{zi}(t)$, from the characteristic equation $\alpha^2+3\alpha+2=0$ or $(\alpha+1)(\alpha+2)=0$, we can get
$$r_{zi}(t) = A_1 e^{-t} + A_2 e^{-2t}$$
substituting the initial conditions into above equation, we obtain
$$\begin{cases} A_1 + A_2 = 1 \\ -A_1 - 2A_2 = 1 \end{cases} \Rightarrow A_1 = 3, \ A_2 = -2$$
so that, the zero input response $r_{zi}(t) = 3e^{-t} - 2e^{-2t}$, $t \geq 0$.

② Determine the impulse response $h(t)$, let $h(t) = [k_1 e^{-t} + k_2 e^{-2t}]u(t)$ (here, $n>m$), substitute it into the differential equation,
$$(k_1+k_2)\delta'(t) - (k_1+2k_2)\delta(t) + k_1 e^{-t}u(t) + 4k_2 e^{-2t}u(t) +$$
$$3(k_1+k_2)\delta(t) - 3k_1 e^{-t}u(t) - 6k_2 e^{-2t}u(t) + 2k_1 e^{-t}u(t) + 2k_2 e^{-2t}u(t)$$
$$= \delta'(t) + \delta(t)$$
that is,
$$\begin{cases} k_1 + k_2 = 1 \\ 2k_1 + k_2 = 1 \end{cases} \text{or} \begin{cases} k_1 = 0 \\ k_2 = 1 \end{cases}$$
therefore, $h(t) = e^{-2t}u(t)$.

③ Find the zero state response $r_{zs}(t)$, from Eq. (2-40),
$$r_{zs}(t) = e(t) * h(t) = \int_{-\infty}^{\infty} (2e^{-\tau} - 1)u(\tau) e^{-2(t-\tau)} u(t-\tau) \mathrm{d}\tau$$
$$= e^{-2t} \int_0^t (2e^{\tau} - e^{2\tau}) \mathrm{d}\tau \cdot u(t)$$

$$= e^{-2t} \left[2e^{\tau} - \frac{1}{2}e^{2\tau} \right]_0^t \cdot u(t)$$

$$= e^{-2t} \left[2e^{t} - 2 - \frac{1}{2}e^{2t} + \frac{1}{2} \right] \cdot u(t)$$

$$= \left[2e^{-t} - \frac{3}{2}e^{-2t} - \frac{1}{2} \right] u(t)$$

④ The complete response

$$r(t) = \left[\underbrace{3e^{-t} - 2e^{-2t}}_{r_{zi}(t)} + \underbrace{2e^{-t} - \frac{3}{2}e^{-2t} - \frac{1}{2}}_{r_{zs}(t)} \right] u(t)$$

$$= \left(\underbrace{5e^{-t} - \frac{7}{2}e^{-2t}}_{\text{transient response}} - \underbrace{\frac{1}{2}}_{\text{steady response}} \right) u(t)$$

2.5 利用 MATLAB 进行连续时间系统的时域分析（Analysis of Continuous-time Systems in Time Domain Based on MATLAB）

2.5.1 冲激响应和阶跃响应的 MATLAB 实现（MATLAB Realization of Unit Impulse Response and Unit Step Response）

MATLAB 为用户提供了专门用于求连续系统的冲激响应和阶跃响应，并绘制其时域波形的函数 impulse( ) 和 step( )。在调用函数 impulse( ) 和 step( ) 时，我们需要用向量来对连续系统进行表示。这里用向量 *a* 和 *b* 来表示微分方程描述的连续系统。两个函数的调用格式分别为：

impulse(b, a)/step(b, a)，以默认方式绘出冲激响应/阶跃响应的时域波形；

impulse(b, a, t) /step(b, a, t)，绘制该系统在 $0 \sim t$ 时间范围内冲激响应/阶跃响应的时域波形；

impulse(b, a, t1: p: t2)/step(b, a, t1: p: t2)，绘制该系统在 $t_1 \sim t_2$ 时间范围内，以时间间隔为 p 均匀取样的冲激响应/阶跃响应的时域波形。

Example 2-18: Consider the differential equation of the LTI continuous-time system $y''(t) + 6y'(t) + 8y(t) = 3x'(t) + 9x(t)$, try to determine the unit impulse response and the unit step response with MATLAB.

Solution: run the following command:
b=[3 9]; a=[1 6 8];
subplot(211)
impulse(b, a);
subplot(212)
step(b, a)

The running result is shown in Figure 2-4.

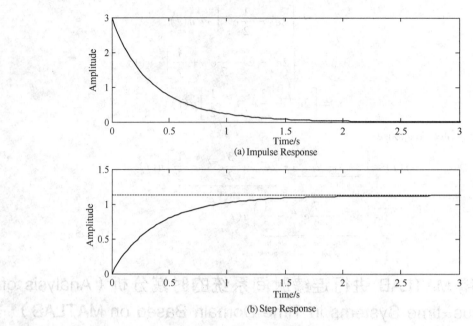

Figure 2-4 the running result of the impulse response and step response

2.5.2 基于 MATLAB 的卷积运算(Convolution Operation Based on MATLAB)

We can realize the convolution operation of two signals by using MATLAB in which the command is $w = conv(u, v)$.

Example 2-19: Assuming that $f_1(t) = u(t)$, $f_2(t) = \delta(t)$, try to realize the convolution operation of $f(t) = f_1(t) * f_2(t)$, $f(t) = f_1(t) * f_1(t)$ and $f(t) = f_2(t) * f_2(t)$ with MATLAB.

Solution: run the following command:

```
%Example 2-19 convolution integral operation
a = 1000;
t1 = -5 : 1/a : 5;
f1 = stepfun(t1, 0);
f2 = stepfun(t1, -1/a) - stepfun(t1, 1/a)
subplot(231)
plot(t1, f1); axis([-5, 5, 0, 1.2]);
title('(a) unit step function f1(t)')
subplot(232); plot(t1, f2);
title('(b) unit impulse signal f2(t)')
y = conv(f1, f2);
r = 2 * length(t1) - 1;
t = -10 : 1/a : 10;
subplot(233);
plot(t, y);
axis([-5, 5, 0, 1.2]);
title('(c) y(t) = f1(t) * f2(t)');
```

```
f11 = conv(f1, f1);
f22 = conv(f2, f2);
subplot(234);
plot(t, f11);
title('(d) f11(t) = f1(t) * f1(t)');
axis([-5, 5, 0, 5000]);
subplot(235);
plot(t, f22);
title('(e) f22(t) = f2(t) * f2(t)')
```

The running results are shown in Figure 2-5.

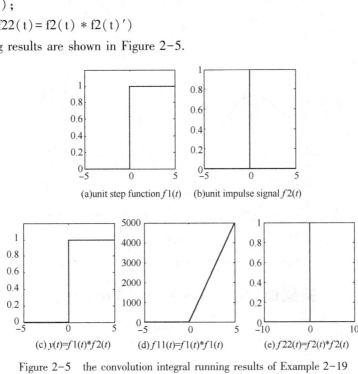

Figure 2-5 the convolution integral running results of Example 2-19

2.5.3 用 MATLAB 求解零状态响应(Solving the Zero State Response Using MATLAB)

MATLAB 的函数 lsim( )能对微分方程描述的 LTI 连续时间系统的响应进行仿真。lsim( )函数能绘制连续系统在指定的任意时间范围内系统响应的时域波形图,还能求出连续系统在指定的任意时间范围内系统响应的数值解,其调用格式为:

lsim(b, a, x, t),绘制零状态响应曲线,x 表示系统的输入信号;

y = lsim(b, a, x, t),不绘制零状态响应曲线,而是求出 t 所定义的时间范围内零状态响应的数值解。

Example 2-20: Consider the differential equation $y''(t) + 8y(t) = x(t)$, the input signal is $x(t) = \cos(t)u(t)$. Try to find the zero state response with MATLAB.

Solution: run the following command:

```
%Example 2-20 finding the zero state response
b = [1]; a = [1 0 1];
```

```
t=0:0.1:10;
x=cos(t);
y=lsim(b, a, x, t);
plot(t, y);
xlabel('time(t)'); ylabel('y(t)'); title('the zero state response')
```
The running resultis shown in Figure 2-6.

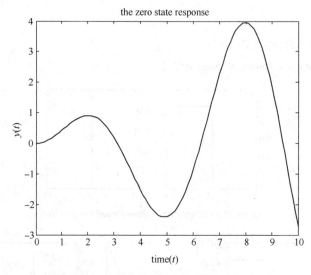

Figure 2-6 the running result of Example 2-20

关键词(Key Words and Phrases)

(1) 时域　　　　　　　　　　time domain
(2) 经典法　　　　　　　　　classical method
(3) 齐次方程　　　　　　　　homogeneous equation
(4) 齐次解　　　　　　　　　homogeneous solution
(5) 特解　　　　　　　　　　particular solution
(6) 线性常系数微分方程　　　linear constant-coefficient differential equation
(7) 激励信号　　　　　　　　exciting signal
(8) 特征方程　　　　　　　　characteristic equation
(9) 特征根　　　　　　　　　characteristic root
(10) 自然频率　　　　　　　　natural frequency
(11) 初始条件　　　　　　　　initial condition
(12) 起始条件　　　　　　　　start condition
(13) 待定系数　　　　　　　　undetermined coefficient
(14) 卷积　　　　　　　　　　convolution
(15) 零输入响应　　　　　　　zero input response
(16) 零状态响应　　　　　　　zero state response
(17) 储能元件　　　　　　　　energy storage element
(18) 跳变值　　　　　　　　　jump value

(19) 冲激平衡法　　　　　　　impulse balance method
(20) 自然(自由)响应　　　　　natural (free) response
(21) 受迫响应　　　　　　　　forced response
(22) 瞬态响应　　　　　　　　transient response
(23) 稳态响应　　　　　　　　steady response
(24) 单位冲激响应　　　　　　unit impulse response
(25) 单位阶跃响应　　　　　　unit step response
(26) 全响应　　　　　　　　　complete response

Exercises

2-1 Solve the following homogeneous differential equations with the specified auxiliary conditions:

(1) $\dfrac{d^2r(t)}{dt^2} + 3\dfrac{dr(t)}{dt} + 2r(t) = 0 \quad r(0) = 0,\ r'(0) = 2$

(2) $\dfrac{d^2r(t)}{dt^2} + 3\dfrac{dr(t)}{dt} + 2r(t) = 0 \quad r(0) = 1,\ r'(0) = -1$

(3) $\dfrac{d^2r(t)}{dt^2} + 2\dfrac{dr(t)}{dt} + r(t) = 0 \quad r(0) = 1,\ r'(0) = 1$

(4) $\dfrac{d^2r(t)}{dt^2} + 2\dfrac{dr(t)}{dt} + 5r(t) = 0 \quad r(0) = 1,\ r'(0) = 1$

2-2 Calculate the convolution of $f_1(t)$ and $f_2(t)$.

(1) $f_1(t) = u(t) - u(t-1),\ f_2(t) = u(t-2) - u(t-3)$

(2) $f_1(t) = u(t),\ f_2(t) = e^{-at}u(t)$

(3) $f_1(t) = \cos\omega_0 t,\ f_2(t) = \delta(t+1) - \delta(t-1)$

(4) $f_1(t) = 2e^{-t}[u(t) - u(t-3)],\ f_2(t) = u(t)$

2-3 Let $f(t) = u(t-3) - u(t-5)$ and $h(t) = e^{-3t}u(t)$

(1) Compute $y(t) = f(t) * h(t)$.

(2) Compute $g(t) = (df(t)/dt) * h(t)$.

(3) How is $g(t)$ related to $y(t)$?

2-4 Suppose that

$$e(t) = \begin{cases} 1, & 0 \leqslant t \leqslant 1 \\ 0, & \text{elsewhere} \end{cases}$$

and $h(t) = e(t/\alpha)$, where $0 < \alpha \leqslant 1$, determine $r(t) = e(t) * h(t)$.

2-5 The following are the differential equations and initial conditions of the systems. Find the zero input response.

(1) $\dfrac{d^2r(t)}{dt^2} + 3\dfrac{dr(t)}{dt} + 2r(t) = 0,\ r(0^-) = 1,\ r'(0^-) = 2$

(2) $\dfrac{d^2r(t)}{dt^2} + 5\dfrac{dr(t)}{dt} + r(t) = 2e'(t) + 5e(t),\ r(0^-) = 1,\ r'(0^-) = 5$

(3) $\dfrac{d^2r(t)}{dt^2} + 4\dfrac{dr(t)}{dt} + 4r(t) = 3e'(t) + 2e(t),\ r(0^-) = -2,\ r'(0^-) = 3$

2-6 Consider the following differential equation
$$\frac{d^2r(t)}{dt^2} + 3\frac{dr(t)}{dt} + 2r(t) = e(t)$$
and $h(0) = 0$, $h'(0) = 1$, calculate the zero state response when the input signal is $e^{-t}u(t)$.

2-7 Determine the jump value $r(0^+)$, for differential equations, the initial conditions and the input signal are given as bellow.

(1) $\dfrac{dr(t)}{dt} + 2r(t) = e(t)$, $r(0^-) = 0$, $e(t) = u(t)$

(2) $\dfrac{dr(t)}{dt} + 2r(t) = 3\dfrac{de(t)}{dt}$, $r(0^-) = 0$, $e(t) = u(t)$

2-8 Consider the differential equation $\dfrac{d^2r(t)}{dt^2} + 5\dfrac{dr(t)}{dt} + 6r(t) = u(t) - u(t-1)$ and suppose that the initial conditions are given as below. Find the complete response of the system respectively.

(1) $r(0^-) = 0$, $r'(0^-) = 0$

(2) $r(0^-) = 0$, $r'(0^-) = 1$

2-9 Determine the complete response and point out the zero input and zero state response, forced and natural response respectively. The differential equations, initial conditions and the exciting signals are given as below.

(1) $\dfrac{d^2r(t)}{at^2} + 3\dfrac{dr(t)}{at} + 2r(t) = \dfrac{de(t)}{dt} + 3e(t)$

$r(0^-) = 1$, $r'(0^-) = 2$, $e(t) = u(t)$

(2) $\dfrac{d^2r(t)}{dt^2} + 2\dfrac{dr(t)}{dt} + r(t) = \dfrac{de(t)}{dt}$

$r(0^-) = 1$, $r'(0^-) = 2$, $e(t) = e^{-t}u(t)$

2-10 Find the impulse response for given differential equations of the systems.

(1) $\dfrac{dr(t)}{dt} + 3r(t) = 2\dfrac{de(t)}{dt}$

(2) $\dfrac{d^2r(t)}{dt^2} + 5\dfrac{dr(t)}{dt} + 6r(t) = \dfrac{d^2e(t)}{dt^2} + 7\dfrac{de(t)}{dt} + 4e(t)$

2-11 The RLC circuit is depicted in the following Figure 2-7, where $L = \dfrac{1}{2}H$, $C = 1F$, $R = \dfrac{1}{3}\Omega$. Let $u_c(t)$ as the output response. Find the impulse and step response respectively.

2-12 Consider the differential equation
$$r''(t) + 3r'(t) + 2r(t) = \frac{1}{2}e'(t) + 3e(t)$$
try to find the step response.

2-13 Consider a system whose input $e(t)$ and output $r(t)$ satisfy the first-order differential equation
$$\frac{dr(t)}{dt} + 2r(t) = e(t)$$

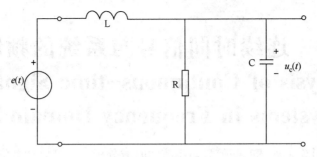

Figure 2-7 RLC circuit

The system also satisfies the condition of initial rest.

(1) Determine the system output $r_1(t)$ when the input is $e_1(t) = e^{3t}u(t)$.

(2) Determine the system output $r_2(t)$ when the input is $e_2(t) = e^{2t}u(t)$.

(3) Determine the system output $r_3(t)$ when the input is $e_3(t) = \alpha e^{3t}u(t) + \beta e^{2t}u(t)$, where α and β are real numbers. Show that $r_3(t) = \alpha r_1(t) + \beta r_2(t)$

2-14 Suppose that the differential equation of a system is given by
$$r''(t) + 5r'(t) + 6r(t) = e^{-t}u(t)$$
find the initial conditions $r(0^-)$ and $r'(0^-)$, and the coefficient c when the complete response is $ce^{-t}u(t)$.

第3章 连续时间信号与系统的频域分析
（Analysis of Continuous-time Signals and Systems in Frequency Domain）

在第2章中，讨论了连续时间信号与系统的时域分析，它是以冲激信号为基本信号，将任意信号分解为一系列加权的冲激信号之和，而系统的零状态响应是输入信号与冲激响应的卷积。本章将以正弦信号或虚指数信号为基本信号，讨论连续时间信号与系统的频域分析。

The representation and analysis of LTI systems through the convolution as developed in Chapter 2 is based on representing signals as linear combinations of shifted impulses. The response of an LTI system to any input is shifted version of the impulse response, leading to the convolution integral. In this Chapter, we explore an alternative representation for signals and LTI systems using complex exponentials. The resulting representations are known as the continuous-time Fourier series and transform.

3.1 周期信号的频域分析（Analysis of Periodic Signals in Frequency Domain）

19世纪初，法国数学家和物理学家傅里叶（Fourier，1768～1830）提出，满足一定条件的时域信号可以表达为一系列正弦（或虚指数）信号的加权叠加，称为信号的傅里叶表达。信号的傅里叶表达揭示了信号的时域与频域之间的内在联系，为信号和系统的分析提供了一种新的方法和途径。

3.1.1 周期信号的频谱分析——傅里叶级数（Spectrum Analysis of Periodic Signals—the Fourier Series）

根据连续Fourier级数（Continuous Fourier Series，CFS）理论，当正交函数集（orthogonal function set）$\{g_i(t), i=1, 2, \cdots\}$中的函数，在区间$(t_0, t_0+T)$上两两正交时，则一个周期为$T$的周期信号，可以展开为$\{g_i(t), i=1, 2, \cdots\}$中各函数的线性组合，这种方法称为周期信号的级数展开（series expansion of periodic signal）。在这些展开式中，最为常见的为傅里叶级数，其中包括三角形式傅里叶级数（trigonometric Fourier series）和指数形式傅里叶级数（exponential Fourier series）。

(1) 三角形式傅里叶级数（trigonometric Fourier series）

不难证明，三角函数集$\{\cos n\omega_0 t, \sin n\omega_0 t, n=0, 1, 2, \cdots\}$在区间$(0, 2\dfrac{\pi}{\omega_0})$是一个完备的正交函数集，此函数集包括无穷多项。这样就可以将一个周期为T的周期信号表示为这个正交函数集中各函数的线性组合。需要指出，这种表示对周期信号$f(t)$有一定要求，即周期信号$f(t)$应满足狄利克雷（Dirichlet）条件：

① 在一个周期内连续或有有限个第一类间断点；
② 在一个周期内函数的极值点是有限的；
③ 在一个周期内函数是绝对可积的，即

$$\int_{t_0}^{t_0+T} |f(t)|\,dt < \infty$$

The Dirichlet conditions are as follows.

Condition 1: In any finite interval of time, there are only a finite number of discontinuities. Furthermore, each of these discontinuities is finite.

Condition 2: In any finite interval of time, $f(t)$ is of bounded variation, that is, there are no more than a finite number of maxima and minima during any single period of the signal.

Condition 3: Over any period, $f(t)$ must be absolutely integrable, that is,

$$\int_{t_0}^{t_0+T} |f(t)|\,dt < \infty$$

因为通常遇到的周期信号大都能满足此条件，所以对此条件以后不再做特别说明。

Under the Dirichlet conditions, any periodic signal with period T can be expanded a linear combination of the functions among trigonometric function set, i.e..

$$\begin{aligned} f(t) &= a_0 + a_1\cos\omega_1 t + a_2\cos2\omega_1 t + \cdots + b_1\sin\omega_1 t + b_2\sin\omega_1 t + \cdots \\ &= a_0 + \sum_{n=1}^{\infty}(a_n\cos n\omega_1 t + b_n\sin n\omega_1 t) \end{aligned} \quad (3-1)$$

Eq. (3-1) is called the trigonometric form of Fourier series expansion. Where n is positive integer, ω_1 is fundamental angel frequency, f_0 is fundamental frequency, and a_0, a_n, b_n are Fourier coefficients.

由正、余弦函数的正交条件，可求得傅里叶系数

$$a_0 = \frac{1}{T}\int_{t_0}^{t_0+T} f(t)\,dt \quad (3-2)$$

$$a_n = \frac{2}{T}\int_{t_0}^{t_0+T} f(t)\cos n\omega_1 t\,dt \quad (3-3)$$

$$b_n = \frac{2}{T}\int_{t_0}^{t_0+T} f(t)\sin n\omega_1 t\,dt \quad (3-4)$$

上述积分中可以任取一个周期，即 t_0 可以任意选取，视计算方便而定。

In Eq. (3-1), the term for a_0 is a constant. The term for $n=1$ is referred to as *the fundamental component*(基波) or *the first harmonic component*(一次谐波). The term for $n=2$ is periodic with half the period (or, equivalently, twice the frequency) of *the fundamental component* and is referred to as *the second harmonic component*(二次谐波). More generally, the component for $n=N$ is referred to as *the Nth harmonic component*(N 次谐波).

若将式(3-1)中的同频率项加以合并，又可以写成三角函数形式的傅里叶级数的另外一种形式

$$\begin{aligned} f(t) &= a_0 + \sum_{n=1}^{\infty}(a_n\cos n\omega_1 t + b_n\sin n\omega_1 t) \\ &= a_0 + \sum_{n=1}^{\infty}\sqrt{a_n^2+b_n^2}\left[\frac{a_n}{\sqrt{a_n^2+b_n^2}}\cos n\omega_1 t - \frac{-b_n}{\sqrt{a_n^2+b_n^2}}\sin n\omega_1 t\right] \\ &= a_0 + \sum_{n=1}^{\infty} c_n(\cos\varphi_n\cos n\omega_1 t - \sin\varphi_n\sin n\omega_1 t) \\ &= c_0 + \sum_{n=1}^{\infty} c_n\cos(n\omega_1 t + \varphi_n) \end{aligned} \quad (3-5)$$

工程上经常将式(3-5)称为余弦形式的傅里叶级数(cosine form of Fourier series)。它表明任何满足狄利克雷条件的周期信号都可以分解为直流分量和一系列谐波分量之和。

这两种三角形式系数的关系为

$$\begin{cases} a_0 = c_0 \\ c_n = \sqrt{a_n^2 + b_n^2} \\ a_n = c_n \cos\varphi_n \\ b_n = -c_n \sin\varphi_n \\ \varphi_n = -\arctan\dfrac{b_n}{a_n} \end{cases} \quad (3\text{-}6)$$

显然，直流分量的大小以及基波与各次谐波的幅度、相位取决于周期信号的波形的具体形式。

在式(3-6)中，各参数 a_n，b_n，c_n 以及 φ_n 都是 n(谐波序号)的函数，也可以说是 $n\omega_1$(谐波频率)的函数。如果以频率为横轴，以幅度或相位为纵轴绘出 c_n 和 φ_n 等的变化关系，便可直观地看出各频率分量的相对大小和相位情况，这样的图分别称为信号的幅度频谱图(magnitude spectrum)和相位频谱图(phase spectrum)。周期信号的频谱只出现在 0，ω_1，$2\omega_1$，…离散的频率点上，这样的频谱叫做离散谱(discrete spectrum)。

Example 3-1：Consider a periodic signal

$$f(t) = 1 + \sqrt{2}\cos\omega_0 t - \cos(2\omega_0 t + \frac{5\pi}{4}) + \sqrt{2}\sin\omega_0 t + \frac{1}{2}\sin 3\omega_0 t ,$$

plot its magnitude and phase spectrum.

Solution：Rewriting $f(t)$ and collecting each of the harmonic components which have the same fundamental frequency, we obtain

$$f(t) = 1 + 2\cos(\omega_0 t - \frac{\pi}{4}) + \cos(2\omega_0 t + \frac{5\pi}{4} - \pi) + \frac{1}{2}\cos(3\omega_0 t - \frac{\pi}{2})$$

$$= 1 + 2\cos(\omega_0 t - \frac{\pi}{4}) + \cos(2\omega_0 t + \frac{\pi}{4}) + \frac{1}{2}\cos(3\omega_0 t - \frac{\pi}{2})$$

we see that,

$$c_0 = 1,\ c_1 = 2,\ c_2 = 1,\ c_3 = \frac{1}{2};\ \varphi_1 = -\frac{\pi}{4},\ \varphi_2 = \frac{\pi}{4},\ \varphi_3 = -\frac{\pi}{2}$$

themagnitude spectrum and phase spectrum are shown in Figure 3-1.

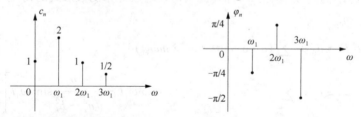

Figure 3-1　the magnitude spectrum and phase spectrum of $f(t)$ in Example 3-1

(2)指数形式傅里叶级数(exponential Fourier series)

满足狄利克雷(Dirichlet)条件的连续周期信号 $f(t)$ 还可以表达为无限项虚指数信号 $e^{jn\omega_1 t}$ 的线性叠加，即

$$f(t) = \sum_{n=-\infty}^{\infty} F(n\omega_1) e^{jn\omega_1 t} \quad (3-7)$$

该式称为周期信号 $f(t)$ 的指数型傅里叶级数(exponential form of Fourier series)，其中 $F(n\omega_1)$ 称为傅里叶系数(Fourier coefficient)，简写为 F_n，利用虚指数信号 $e^{jn\omega_1 t}$ 的正交性，可以很容易地确定 $F(n\omega_1)$，即

$$F(n\omega_1) = \frac{1}{T}\int_{t_0}^{t_0+T} f(t) e^{-jn\omega_1 t} dt \quad (3-8)$$

Given a continuous-time periodic signal $f(t)$, if it satisfies the Dirichlet conditions, then it has a Fourier series representation given by Eq. (3-7) and Eq. (3-8), where F_n is complex value. This pair of equations defines the Fourier series of a periodic continuous-time signal. The set of coefficients $\{F_n\}$ are often called the Fourier series coefficients or the spectral coefficients of $f(t)$.

一般来说，F_n 是复常数，可以表示成模和幅角的形式

$$F_n = |F_n| e^{j\varphi_n} \quad (3-9)$$

在三角函数标准形式中，c_n 是第 n 次谐波分量的振幅，但在指数形式中，F_n 要与相对应的第 $-n$ 项 F_{-n} 合并，构成第 n 次谐波分量的振幅和相位。

根据欧拉公式，式(3-8)可写为

$$F_n = \frac{1}{T}\int_{t_0}^{t_0+T} f(t) e^{-jn\omega_1 t} dt = \frac{1}{T}\int_{t_0}^{t_0+T} f(t)(\cos n\omega_1 t - j\sin n\omega_1 t) dt$$

$$= \frac{1}{2}(a_n - jb_n)$$

因此，可得指数形式与三角形式傅里叶系数之间的关系为

$$\begin{cases} F_0 = a_0 = c_0 \\ |F_n| = \frac{1}{2}c_n, \quad n \neq 0 \\ \varphi_n = -\arctan\dfrac{b_n}{a_n} \end{cases} \quad (3-10)$$

Example 3-2: For Example 3-1, write its exponential from of Fourier series by using the Enler's formula.

Solution:

$$f(t) = 1 + \sqrt{2}\cos\omega_0 t - \cos\left(2\omega_0 t + \frac{5\pi}{4}\right) + \sqrt{2}\sin\omega_0 t + \frac{1}{2}\sin 3\omega_0 t$$

$$= 1 + \frac{\sqrt{2}}{2}[e^{j\omega_0 t} + e^{-j\omega_0 t}] - 0.5[e^{j(2\omega_0 t + 5\pi/4)} + e^{-j(2\omega_0 t + 5\pi/4)}]$$

$$- \frac{\sqrt{2}}{2}j[e^{j\omega_0 t} - e^{-j\omega_0 t}] - 0.25j[e^{j3\omega_0 t} - e^{-j3\omega_0 t}]$$

$$= 1 + \left(\frac{\sqrt{2}}{2} + \frac{\sqrt{2}}{2}j\right) e^{-j\omega_0 t} + \left(\frac{\sqrt{2}}{2} - \frac{\sqrt{2}}{2}j\right) e^{j\omega_0 t}$$

$$- 0.5e^{-j\frac{5\pi}{4}} e^{-j2\omega_0 t} - 0.5e^{j\frac{\pi}{4}} e^{j2\omega_0 t} + 0.25j e^{-j3\omega_0 t} - 0.25j e^{j3\omega_0 t}$$

we see that,

$$F_0 = 1$$
$$F_1 = \frac{\sqrt{2}}{2} - \frac{\sqrt{2}}{2}j, \quad F_{-1} = \frac{\sqrt{2}}{2} + \frac{\sqrt{2}}{2}j$$
$$F_2 = -0.5e^{j\frac{\pi}{4}} = -\frac{\sqrt{2}}{4}(1+j), \quad F_{-2} = -0.5e^{-j\frac{\pi}{4}} = -\frac{\sqrt{2}}{4}(1-j)$$
$$F_3 = -0.25j, \quad F_{-3} = 0.25j$$

In Figure 3-2, we show a bar graph of the magnitude and phase of F_n.

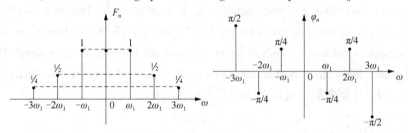

Figure 3-2　the bar graph of the magnitude and phase of F_n in Example 3-2

可以看出，同一个信号，既可以展开成三角形式的傅里叶级数，又可以展开成指数形式的傅里叶级数，二者形式虽然不同，实质是完全一致的。指数形式展开式中有负频率(negative frequency)，这只是表示形式的问题，并不表示真正有负频率的分量，负的频率项与相应的正频率项合并起来才代表一个振荡分量。在双边频谱图(bilateral spectrum)上，$n\omega_1$ 频率分量的振幅一分为二，在 $-n\omega_1$ 和 $n\omega_1$ 的频率上各据一半（$|F_{-n}|=|F_n|=c_n/2$），单独考虑正频率项或单独考虑负频率项都是不全面的。

最后指出，当 $f(t)$ 是实信号(real signal)时，有下述公式：

$$\begin{cases} |F_{-n}| = |F_n| = \dfrac{\sqrt{a_n^2 + b_n^2}}{2} \\ \varphi_{-n} = -\varphi_n = \arctan \dfrac{b_n}{a_n} \end{cases} \quad (3-11)$$

即其复振幅的模是 n 或是频率 $n\omega_1$ 的偶函数(even function)，而复振幅的幅角(初相位)是 n 或是频率 $n\omega_1$ 的奇函数(odd function)，反映到谱线图上，前者的谱线关于纵轴对称，而后者的谱线关于原点对称。

3.1.2　傅里叶级数系数与函数对称性的关系(The Relationship Between Fourier Series Coefficients and the Symmetry of Function)

当周期信号的波形具有某种对称性(symmetry)时，其相应的傅里叶级数的系数会呈现出某些特征。周期信号的对称性大致分为两类：一类是波形关于原点或纵轴对称，即我们所熟悉的奇函数、偶函数，这种对称性决定了展开式中是否有正弦项或余弦项；另一类是波形前半周期与后半周期是否相同或成镜像对称关系(半波对称，half-wave symmetry)，这种对称性决定了展开式中是否含有偶次或奇次谐波(even or odd harmonic)。下面具体讨论对称条件对傅里叶级数系数的影响。

(1) 偶函数(even function)

For even function, we have $f(t)=f(-t)$, then in Eq. (3-3) and (3-4), $f(t)\cos n\omega_0 t$ is an even function, and $f(t)\sin n\omega_0 t$ is an odd function. Therefore,

$$\begin{cases} a_n = \dfrac{2}{T}\int_{-\frac{T}{2}}^{\frac{T}{2}} f(t)\cos n\omega_1 t\mathrm{d}t = \dfrac{4}{T}\int_0^{\frac{T}{2}} f(t)\cos n\omega_1 t\mathrm{d}t \\ b_n = \dfrac{2}{T}\int_{-\frac{T}{2}}^{\frac{T}{2}} f(t)\sin n\omega_1 t\mathrm{d}t = 0 \end{cases} \quad (3-12)$$

It is visible that the Fourier series of any even function is composed of only a constant and cosine function.

偶函数的傅里叶级数不含正弦项，只含有余弦项和直流项。

(2) 奇函数(odd function)

For odd function, we have $f(t) = -f(-t)$, then in Eq. (3-3) and (3-4), $f(t)\cos n\omega_0 t$ is an odd function, and $f(t)\sin n\omega_0 t$ is an even function. Therefore,

$$\begin{cases} a_n = \dfrac{2}{T}\int_{-\frac{T}{2}}^{\frac{T}{2}} f(t)\cos n\omega_1 t\mathrm{d}t = 0 \\ b_n = \dfrac{2}{T}\int_{-\frac{T}{2}}^{\frac{T}{2}} f(t)\sin n\omega_1 t\mathrm{d}t = \dfrac{4}{T}\int_0^{\frac{T}{2}} f(t)\sin n\omega_1 t\mathrm{d}t \end{cases} \quad (3-13)$$

Fourier series expansion of odd function does not contain the constant and cosine function, only containing the sinusoidal term.

奇函数的傅里叶级数展开式中不含直流项和余弦项，只含有正弦项。

(3) 半波镜像函数(half-wave mirror function)

如果函数波形沿时间轴平移半个周期并上下翻转后得到的波形与原波形重合，即

$$f(t) = -f\left(t \pm \dfrac{T}{2}\right) \quad (3-14)$$

则称此函数为半波镜像函数。图 3-3 是半波镜像周期函数的一个实例。

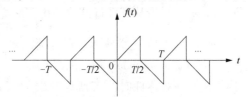

图 3-3　半波镜像周期函数

由式(3-3)得

$$a_n = \dfrac{2}{T}\int_{-\frac{T}{2}}^{0} f(t)\cos n\omega_1 t\mathrm{d}t + \dfrac{2}{T}\int_0^{\frac{T}{2}} f(t)\cos n\omega_1 t\mathrm{d}t \quad (3-15)$$

将 $f(t) = -f\left(t + \dfrac{T}{2}\right)$ 代入第一个积分中，有

$$\begin{aligned}
&\dfrac{2}{T}\int_{-\frac{T}{2}}^{0}\left[-f\left(t+\dfrac{T}{2}\right)\cos n\omega_1\left(t+\dfrac{T}{2}-\dfrac{T}{2}\right)\right]\mathrm{d}t \\
&= \dfrac{2}{T}\int_{-\frac{T}{2}}^{0}\left[-f\left(t+\dfrac{T}{2}\right)\cos n\omega_1\left(t+\dfrac{T}{2}\right)\cos n\pi\right]\mathrm{d}t \\
&= -\cos n\pi \dfrac{2}{T}\int_0^{\frac{T}{2}} f(t)\cos n\omega_1 t\mathrm{d}t
\end{aligned} \quad (3-16)$$

将式(3-16)代入到式(3-15)，得

$$a_n = \frac{2}{T}\int_0^{\frac{T}{2}} f(t)\cos n\omega_1 t\, dt - \cos n\pi \frac{2}{T}\int_0^{\frac{T}{2}} f(t)\cos n\omega_1 t\, dt$$

$$= (1 - \cos n\pi)\frac{2}{T}\int_0^{\frac{T}{2}} f(t)\cos n\omega_1 t\, dt \tag{3-17}$$

$$= \begin{cases} 0, & n\text{ 为偶数} \\ \dfrac{4}{T}\int_0^{\frac{T}{2}} f(t)\cos n\omega_1 t\, dt, & n\text{ 为奇数} \end{cases}$$

同理可得

$$b_n = \begin{cases} 0, & n\text{ 为偶数} \\ \dfrac{4}{T}\int_0^{\frac{T}{2}} f(t)\sin n\omega_1 t\, dt, & n\text{ 为奇数} \end{cases} \tag{3-18}$$

可以看出，半波镜像函数的傅里叶级数展开式中只含奇次谐波而不含偶次谐波项，故又称为奇谐函数。

Fourier series expansionof half-wave mirror function contains only odd harmonics. So half-wave mirror function is also called *odd harmonic function*.

（4）半波重叠函数（half-wave overlapping function）

如果函数波形沿时间轴平移半个周期后得到的波形与原波形重合，即

$$f(t) = f\left(t \pm \frac{T}{2}\right) \tag{3-19}$$

则称此函数为半波重叠函数。图 3-4 是半波重叠周期函数的一个实例。

这种函数的周期实际上是 $\dfrac{T}{2}$，即其基频实际上是 $\dfrac{2\pi}{T/2} = 2 \cdot \dfrac{2\pi}{T} = 2\omega_1$。按以 T 为周期、以 $\omega_1 = \dfrac{2\pi}{T}$ 为基频进行谐波分析，当然就不会存在奇次谐波项（所有谐波频率都是 $2\omega_1$ 的整数倍，即 ω_1 的偶数倍），是偶谐函数（*even harmonic function*）。

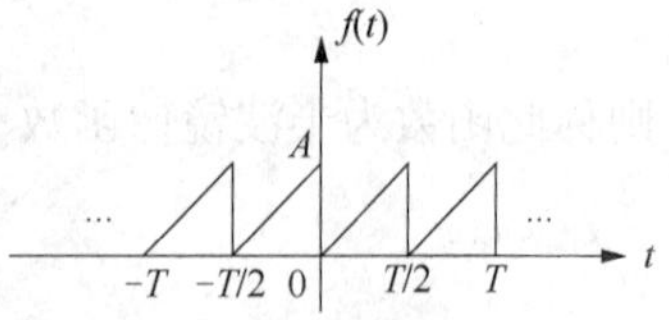

图 3-4 半波重叠周期函数

综上所述，当信号波形具有某种对称性时，其傅里叶级数展开式中有些项就不出现。掌握这一特点，可对信号包含哪些谐波成分做出迅速的判断，从而简化展开式系数的计算。

In summary, somecomponents in the Fourier series expansion will not appear when the signal has a certain symmetry. According to the characteristic, we can make quick judgments which kind of harmonic component that's contained in the signal, so as to simplify the calculation of Fourier series coefficients.

3.1.3 周期矩形脉冲信号的傅里叶级数（Fourier Series of Periodic Rectangular Signal）

周期矩形脉冲是典型的周期信号，其频谱函数具有周期信号频谱的基本特点。通过分析周期矩形脉冲的频谱，可以了解周期信号频谱的一般规律。

设周期矩形脉冲 $f(t)$ 脉冲宽度为 τ，脉冲幅度为 E，周期为 T，波形如图 3-5 所示。

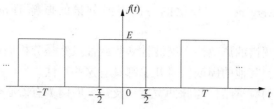

图 3-5 周期矩形脉冲信号

该信号在第一个周期内的表达式为

$$f(t) = \begin{cases} E, & -\dfrac{\tau}{2} < t < \dfrac{\tau}{2} \\ 0, & \text{其他} \end{cases} \tag{3-20}$$

The periodic rectangular signals is plotted in Figure 3-5. Specifically, over one period, the representation of $f(t)$ is shown in Eq. (3-20) and periodically repeats with period T. Where, the width of the signal is τ, and the amplitude is E.

将 $f(t)$ 展开成指数形式傅里叶级数，由式(3-8)可得

$$\begin{aligned} F(n\omega_1) &= \dfrac{1}{T} \int_{-\frac{\tau}{2}}^{\frac{\tau}{2}} E e^{-jn\omega_1 t} dt \\ &= \dfrac{E}{T} \dfrac{1}{-jn\omega_1} e^{-jn\omega_1} \bigg|_{-\frac{\tau}{2}}^{\frac{\tau}{2}} = \dfrac{E}{T} \dfrac{2}{n\omega_1} \dfrac{1}{2j}(e^{j\frac{n\omega_1 \tau}{2}} - e^{-j\frac{n\omega_1 \tau}{2}}) \\ &= \dfrac{2E}{Tn\omega_1} \sin\dfrac{n\omega_1 \tau}{2} \\ &= \dfrac{E\tau}{T} Sa\left(\dfrac{n\omega_1 \tau}{2}\right) \end{aligned} \tag{3-21}$$

可见，本例中的 $F(n\omega_1)$ 为 $n\omega_1$ 的实函数，图形与 $Sa(t)$ 的图形类似，即 $Sa(t)$ 是频谱各分量的包络线(envelope)。周期矩形脉冲信号的频谱如图 3-6 所示。其中第一个包络零点为 $\omega = \dfrac{2\pi}{\tau}$。

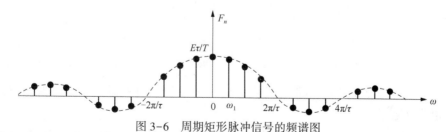

图 3-6 周期矩形脉冲信号的频谱图

In this case, $F(n\omega_1)$ is a real function of $n\omega_1$, and the bar graph of these coefficients are shown in Figure 3-6. Obviously, the graph is similar to that of $Sa(t)$, that is, the envelope of the amplitude of the harmonic components in the periodic square wave is the sinc function $Sa(t)$.

由图 3-6 可以看出，此周期信号频谱具有以下几个特点：

第一，**离散性(discrete)**：频谱由不连续的谱线组成，每一条谱线代表一个正弦分量，所以此谱称为不连续谱或离散谱。

第二，**谐波性(harmonic)**：频谱的每一条谱线只能出现在基波频率 ω_1 的整数倍频率上。

第三，收敛性(convergence)：频谱的各次谐波分量的振幅随 $n\omega_1$ 的增大而减小，当 $n\omega_1 \to \infty$ 时，$|F_n| \to 0$。

上述关于周期信号频谱的离散性、谐波性和收敛性虽然是分析周期矩形脉冲信号而得到的，但它具有普遍意义，其他的周期信号几乎都具有这些特性。

周期矩形脉冲信号的频谱结构与脉冲宽度 τ 及信号周期 T 有着必然的联系。当周期 T 为定值时，其基波频率 $\omega_1 = \dfrac{2\pi}{T}$ 为一确定值，而随着 τ 的减小，其第一个包络零点增大，而各次谐波分量的振幅同时减小。当脉冲宽度 τ 为定值时，其频谱包络的第一个零点为一确定值。随着周期 T 的增大，基波频率 $\omega_1 = \dfrac{2\pi}{T}$ 逐渐减小，谱线变密，而各次谐波分量的振幅也同时减小。

不难看出，当周期 T 无限增大时，频谱的谱线无限密集，各次谐波分量的振幅趋于无穷小量，此时周期信号将趋于单脉冲的非周期信号。有关非周期信号的频谱将在本章 3.2 节讨论。

From Figure 3-6, we see that as T increases, or equivalently, as the fundamental frequency $\omega_1 = \dfrac{2\pi}{T}$ decreases, the envelope is sampled with a closer and closer spacing. As T becomes arbitrarily large, the original periodic square wave approaches a rectangular pulse (i.e., all that remains in the time domain is an aperiodic signal corresponding to one period of the square wave). The spectrum of aperiodic signal will be discussed in Section 3.2.

周期矩形脉冲信号含有无穷多条谱线，也就是说，周期矩形脉冲信号可表示为无穷多个正弦分量之和。在信号的传输过程中，要求一个传输系统能将这无穷多个正弦分量不失真地传输显然是不可能的。实际工作中，应要求传输系统能将信号中的主要频率分量传输过去，以满足失真方面的要求。周期矩形脉冲信号的主要能量集中在第一个零点之内，因而，常常将 $\omega = 0 \sim \dfrac{2\pi}{\tau}$ 这段频率范围称为矩形脉冲信号的频带宽度(frequency bandwidth)。记为

$$B_\omega = \dfrac{2\pi}{\tau} \quad \text{rad/s} \tag{3-22}$$

或

$$B_f = \dfrac{1}{\tau} \quad \text{Hz} \tag{3-23}$$

显然，信号的频带宽度 B_f 与信号持续时间 τ 成反比，也就是说，信号持续时间越短，该信号的频带越宽。

Obviously, the frequency bandwidth of the signal is inversely proportional to the duration of the signal, that is to say, the shorter the duration of the signal is, the wider the frequency bandwidth of the signal is.

3.2 非周期信号的傅里叶变换(Fourier Transform of Aperiodic Signals)

从时域可以看到，如果一个周期信号的周期趋于无穷，则周期信号将演变成一个非周期

信号。把非周期信号看成是周期信号在周期趋于无穷时的极限,从而考察连续时间傅里叶级数在 T 趋于无穷时的变化,就能得到对非周期信号的频域表示方法。

3.2.1 非周期信号的频域表示——傅里叶变换（Frequency Domain Representation of Aperiodic Signals—Fourier Transform）

在上一节关于周期信号的傅里叶级数的讨论中,当周期矩形脉冲信号的周期 T 趋于无限大时,周期信号就转化为非周期的单脉冲信号。当周期信号的周期 T 趋于无穷大时,其对应频谱的谱线间隔 $\omega_1 = \dfrac{2\pi}{T}$ 趋于无限小,这样离散频谱就变成了连续频谱,同时构成信号的正弦分量的振幅也趋于无限小量。由此可见,对非周期信号采用傅里叶级数的分析方法显然是不行的。虽然组成非周期信号的各正弦分量的振幅趋于无穷小量,但这并不意味着非周期信号不含正弦分量。从物理概念上考虑,既然成为一个信号,必然含有一定的能量,无论信号怎样分解,其所含能量是不变的。

From last section, we have seen that as T increases, or equivalently, as the fundamental frequency $\omega_1 = \dfrac{2\pi}{T}$ decreases, the envelope is sampled with a closer and closer spacing. Specifically, we think of an aperiodic signal as the limit of a periodic signal as the period becomes infinity. As the period becomes infinite, the frequency components form a continuum and the Fourier series sum becomes an integral. We examine the limiting behavior of the Fourier series representation for this signal.

基于上述原因,对非周期信号,不能采用傅里叶级数的复振幅来表示其频谱,必须引入一个新的量——频谱密度函数(spectrum density function)。下面由周期信号的傅里叶级数推导出傅里叶变换,从而引出频谱密度函数的概念。

Based on the above reasons, we can't represent the spectrum of the aperiodic signal with Fourier series, and we must introduce a new concept——spectrum density function.

设有一周期信号 $f(t)$,其指数形式的傅里叶级数为

$$f(t) = \sum_{-\infty}^{\infty} F(n\omega_1) e^{jn\omega_1 t} \tag{3-24}$$

其复振幅为

$$F(n\omega_1) = \frac{1}{T} \int_{-\frac{T}{2}}^{\frac{T}{2}} f(t) e^{-jn\omega_1 t} dt \tag{3-25}$$

当 $T \to \infty$ 时,$|F(n\omega_1)|$ 趋于无穷小量。若给上式两端同乘以 T,则有

$$T \cdot F(n\omega_1) = \frac{F(n\omega_1)}{f} = \int_{-\frac{T}{2}}^{\frac{T}{2}} f(t) e^{-jn\omega_1 t} dt \tag{3-26}$$

对于非周期信号,$T \to \infty$,谱线间隔 ω_1 趋于无穷小量 $d\omega$,离散频率 $n\omega_1$ 变成连续频率 ω。在这种极限情况下,$\dfrac{F(n\omega_1)}{f}$ 趋于有限值,且为一连续函数,通常记为 $F(\omega)$,即

$$F(\omega) = \lim_{T \to \infty} T \cdot F(n\omega_1) = \lim_{T \to \infty} \int_{-\frac{T}{2}}^{\frac{T}{2}} f(t) e^{-jn\omega_1 t} dt = \int_{-\infty}^{\infty} f(t) e^{-j\omega t} dt \tag{3-27}$$

$F(\omega)$ 称为非周期信号 $f(t)$ 的频谱密度函数。

Suppose that there is a periodic signal $f(t)$, and its exponential Fourier series is shown in Eq. (3-24), and the Fourier series coefficients $F(n\omega_1)$ is shown in Eq. (3-25). As $T \to \infty$, $|F(n\omega_1)| \to 0$, and $\dfrac{F(n\omega_1)}{f}$ is a finite value, the limit in Eq. (3-27) becomes a representation of $F(\omega)$, which is referred to as the spectrum density function.

同样，对式(3-24)也可改写为如下形式：

$$f(t) = \sum_{-\infty}^{\infty} F(n\omega_1) e^{jn\omega_1 t} = \sum_{-\infty}^{\infty} \frac{F(n\omega_1)}{\omega_1} \cdot \omega_1 \cdot e^{jn\omega_1 t}$$

上式在 $T\to\infty$ 时，$\dfrac{F(n\omega_1)}{\omega_1} \to \dfrac{F(\omega)}{2\pi}$，$\omega_1 \to d\omega$，$\Sigma$ 转化为从 $-\infty$ 到 $+\infty$ 的积分，从而得到

$$f(t) = \lim_{T\to\infty} \sum_{-\infty}^{\infty} \frac{T \cdot F(n\omega_1)}{2\pi} \cdot e^{jn\omega_1 t} \cdot \omega_1 = \frac{1}{2\pi} \int_{-\infty}^{\infty} F(\omega) e^{j\omega t} d\omega \tag{3-28}$$

式(3-27)和式(3-28)为非周期信号的频谱表达式，称其为傅里叶变换。

Eq. (3-27) and (3-28) are collectively referred as Fourier transform pair. Eq. (3-27) represents Fourier transform, while Eq. (3-28) represents inverse Fourier transform. Usually, use the following symbol to express.

Fourier transform pair：

$$\begin{cases} F(\omega) = \mathscr{F}[f(t)] = \int_{-\infty}^{\infty} f(t) e^{-j\omega t} dt \\ f(t) = \mathscr{F}^{-1}[F(\omega)] = \dfrac{1}{2\pi} \int_{-\infty}^{\infty} F(\omega) e^{j\omega t} d\omega \end{cases} \tag{3-29}$$

The corresponding relationship between $f(t)$ and $F(\omega)$ is denoted by

$$f(t) \leftrightarrow F(\omega)$$

$F(\omega)$ 一般为复数，可以写作

$$F(\omega) = |F(\omega)| e^{j\varphi(\omega)}$$

式中，$|F(\omega)|$ 代表各频率分量的相对大小；$\varphi(\omega)$ 是 $F(\omega)$ 的相位函数，它表示信号中各频率分量之间的相位关系。习惯上把 $|F(\omega)| \sim \omega$ 曲线和 $\varphi(\omega) \sim \omega$ 曲线分别称为幅度频谱(amplitude spectrum)和相位频谱(phase spectrum)。

非周期信号和周期信号一样，也可以分解为无限多个频率为 ω（ω 从 $-\infty$ 到 $+\infty$ 连续变化）、复振幅为 $\dfrac{F(\omega)}{2\pi}$ 的指数分量 $e^{j\omega t}$ 的连续和(积分)。

The aperiodic signals can still be represented as a linear combination of complex exponentials $e^{j\omega t}$. The magnitude of component with frequency ω is $\dfrac{F(\omega)}{2\pi}$.

需要指出，在上面推导傅里叶变换时并未遵循数学上的严格步骤。从理论上讲，$f(t)$ 应满足一定的条件才可存在傅里叶变换。一般来说，傅里叶变换存在的充分条件(sufficient condition)为 $f(t)$ 满足绝对可积(absolutely integrable)，即

$$\int_{-\infty}^{\infty} |f(t)| dt < \infty \tag{3-30}$$

但这并不是必要条件(necessary condition)，后面将看到引入广义函数(generalized function)的概念之后，许多并不满足绝对可积条件的信号也存在傅里叶变换。

3.2.2 典型非周期信号的傅里叶变换(Fourier Transform of Typical Aperiodic Signals)

本节利用傅里叶变换来分析几种典型非周期信号的频谱。

(1) 矩形脉冲信号(rectangular pulse signal)

已知矩形脉冲信号的表达式为

$$f(t) = E\left[u\left(t+\frac{\tau}{2}\right) - u\left(t-\frac{\tau}{2}\right)\right] \tag{3-31}$$

波形如图 3-7(a)所示。其中 E 为脉冲幅度，τ 为脉冲宽度，也可记为 $G_\tau(t)$。

$$\begin{aligned}F(\omega) &= \int_{-\infty}^{\infty} f(t) e^{-j\omega t} dt \\ &= E\int_{-\frac{\tau}{2}}^{\frac{\tau}{2}} \cos\omega t\, dt \\ &= \frac{2E}{\omega}\sin\omega t\Big|_0^{\frac{\tau}{2}} \\ &= \frac{2E}{\omega}\sin\frac{\omega\tau}{2} = E\tau Sa\left(\frac{\omega\tau}{2}\right)\end{aligned} \tag{3-32}$$

因为 $F(\omega)$ 在这里是实函数，通常用一条 $F(\omega)$ 曲线同时表示幅度频谱和相位频谱，如图 3-7(b)所示。

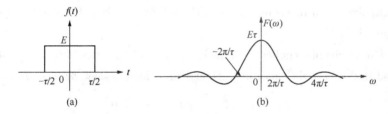

图 3-7 矩形脉冲信号的波形和频谱

Here, because $F(\omega)$ is a real function, we use a curve shown in Figure 3-7(b) to express the amplitude spectrum and phase spectrum at the same time.

(2) 单边指数信号(unilateral exponential signal)

已知单边指数信号的表达式为

$$f(t) = e^{-at}u(t), \quad a > 0 \tag{3-33}$$

其波形如图 3-8 所示。

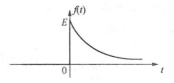

图 3-8 单边指数信号的波形

傅里叶变换为

$$F(\omega) = \int_0^\infty e^{-at} e^{-j\omega t} dt = \int_0^\infty e^{-(a+j\omega)t} dt$$

$$= -\frac{1}{a+j\omega} e^{-(a+j\omega)t} \Big|_0^\infty \qquad (3-34)$$

$$= \frac{1}{a+j\omega}$$

其幅度频谱和相位频谱分别为

$$|F(\omega)| = \frac{1}{\sqrt{a^2+\omega^2}} \qquad (3-35)$$

$$\varphi(\omega) = -\arctan\left(\frac{\omega}{a}\right) \qquad (3-36)$$

波形如图 3-9 所示。

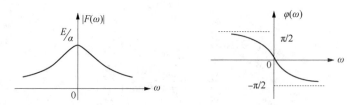

图 3-9 单边指数信号的幅度和相位频谱

Since this Fourier transform is complex valued, to plot it as a function of ω, we express $F(\omega)$ in terms of its magnitude and phase in Eq. (3-35) and Eq. (3-36). Each of these components is sketched in Figure 3-9.

Note that if a is complex rather than real, then $f(t)$ is absolutely integrable as long as $(Re-a) > 0$, and in this case the preceding calculation yields the same form for $F(\omega)$. That is,

$$F(\omega) = \frac{1}{a+j\omega}, \quad (Re-a) > 0$$

(3) 双边指数信号(bilateral exponential signal)

已知双边指数信号的表达式为

$$f(t) = e^{-a|t|}, \quad a > 0 \qquad (3-37)$$

其傅里叶变换为

$$F(\omega) = \int_{-\infty}^{\infty} e^{-a|t|} e^{-j\omega t} dt = \int_{-\infty}^{0} e^{at} e^{-j\omega t} dt + \int_0^\infty e^{-at} e^{-j\omega t} dt$$

$$= \frac{1}{a-j\omega} + \frac{1}{a+j\omega} = \frac{2a}{a^2+\omega^2} \qquad (3-38)$$

双边指数信号的波形及频谱图如图 3-10 所示。

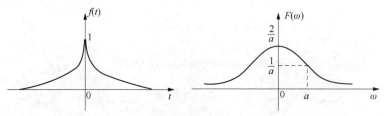

图 3-10 双边指数信号的波形及其频谱

(4) 单位冲激信号(unit impulse signal)

单位冲激信号 $\delta(t)$ 的傅里叶变换是

$$F(\omega) = \int_{-\infty}^{\infty} \delta(t) e^{-j\omega t} dt = 1 \qquad (3-39)$$

可见，冲激函数 $\delta(t)$ 的频谱是常数1，也就是说，$\delta(t)$ 中包含了所有的频率分量，而各频率分量的频谱密度都相等。这种频谱常称为"均色谱"或"白色谱"

The result implied that impulse contains unity contributions from complex sinusoids of all frequencies, from $\omega = 0$ to $\omega = \infty$. $\delta(t)$ and it's Fourier transform are illustrated in Figure 3-11.

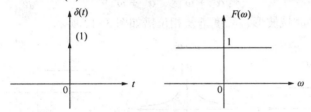

Figure 3-11 the unit impulse signal and it's Fourier transform

(5) 单位阶跃信号(unit step signal)

因为单位阶跃信号 $u(t)$ 不满足绝对可积条件，不能根据傅里叶变换的定义直接来求，但它却存在傅里叶变换。

Because the unit step signal $u(t)$ doesn't meet the absolutely integrable condition, so the Fourier transform can't be find according to the definition, but it certainly has the Fourier transform.

We can use the following limitation form to express $u(t)$

$$u(t) = \lim_{a \to 0} e^{-at} \qquad t > 0$$

this is the limitation of unilateral exponential signal, using the previous result and exchanging the order of the limitation and the integral, we get

$$F(\omega) = \int_{-\infty}^{\infty} \lim_{a \to 0} e^{-at} e^{-j\omega t} dt = \lim_{a \to 0} \int_{-\infty}^{\infty} e^{-at} e^{-j\omega t} dt$$

$$= \lim_{a \to 0} \frac{1}{a + j\omega}$$

$$= \lim_{a \to 0} \left[\frac{a}{a^2 + \omega^2} - j \frac{\omega}{a^2 + \omega^2} \right]$$

Consider the limitation of the first term

$$\lim_{a \to 0} \frac{a}{a^2 + \omega^2} = \begin{cases} \infty & \omega = 0 \\ 0 & \omega \neq 0 \end{cases}$$

obviously, the limitation is an impulse function $\delta(\omega)$, and the impulse strength is

$$\lim_{a \to 0} \int_{-\infty}^{\infty} \frac{a}{a^2 + \omega^2} d\omega = \lim_{a \to 0} \int_{-\infty}^{\infty} \frac{\frac{1}{a}}{1 + \left(\frac{\omega}{a}\right)^2} d\omega$$

$$= \lim_{a \to 0} \int_{-\infty}^{\infty} \frac{d\frac{\omega}{a}}{1 + \frac{\omega^2}{a^2}} = \lim_{a \to 0} \arctan \frac{\omega}{a} \Big|_{-\infty}^{\infty} = \pi$$

thus, the limitation of the first term is $\lim\limits_{\alpha \to 0} \dfrac{a}{\alpha^2 + \omega^2} = \pi\delta(\omega)$

Then, we consider the limitation of the second term

$$\lim_{\alpha \to 0} \dfrac{-j\omega}{\alpha^2 + \omega^2} = \dfrac{1}{j\omega}$$

therefore, the Fourier transform of the unit step signal is

$$F(\omega) = \lim_{\alpha \to 0}\left[\dfrac{\alpha}{\alpha^2 + \omega^2} - j\dfrac{\omega}{\alpha^2 + \omega^2}\right] = \pi\delta(\omega) + \dfrac{1}{j\omega} \tag{3-40}$$

单位阶跃信号的时域波形、幅度谱及相位谱如图 3-12 所示。

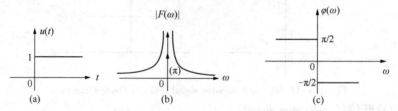

图 3-12　单位阶跃信号的波形及其幅度谱、相位谱

(6) 直流信号(DC signal)

直流信号可以表示为

$$f(t) = 1$$

按照傅里叶变换的定义无法求出这个积分。比照单位冲激信号的傅里叶变换，由傅里叶逆变换容易求得。

According to the definition of the Fourier transform, we cannot find out the integral. But contrasting with the Fourier transform of unit impulse signal, the integral can be easily obtained by the definition of the inverse of Fourier transform.

$$\delta(t) = \dfrac{1}{2\pi}\int_{-\infty}^{+\infty} 1 \cdot e^{j\omega t}d\omega$$

令 t 和 ω 互换，即 $t \leftrightarrow \omega$，

$$\delta(\omega) = \dfrac{1}{2\pi}\int_{-\infty}^{+\infty} 1 \cdot e^{j\omega t}dt$$

令 $\omega \leftrightarrow -\omega$，并且 $\delta(-\omega) = \delta(\omega)$，得到

$$\mathscr{F}[1] = \int_{-\infty}^{+\infty} 1 \cdot e^{-j\omega t}dt = 2\pi\delta(\omega) \tag{3-41}$$

此外，直流信号也可用求双边指数信号的频谱中使 $a \to 0$ 的极限来求得直流信号的频谱。直流信号及其频谱如图 3-13 所示。

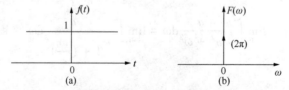

图 3-13　单位阶跃信号的波形及其幅度谱、相位谱

In addition, the spectrum of DC signal can also be obtained from the limit of the spectrum of bilateral exponential signal when $a \to 0$. DC signal and its spectrum are shown in Figure 3-13.

(7) 符号函数(sign function)

符号函数也称正负函数,属于奇异函数,记为 sgn(t)。

The sign function also calls positive and negative function, which belongs to the singular function, denoted as sgn(t).

$$\text{sgn}(t) = \begin{cases} 1, & t > 0 \\ -1, & t < 0 \end{cases}$$

Because sgn(t) = $2u(t) - 1$, so that

$$\mathscr{F}[\text{sgn}(t)] = \int_{-\infty}^{\infty} \text{sgn}(t) e^{-j\omega t} dt = \int_{-\infty}^{\infty} [2u(t) - 1] e^{-j\omega t} dt$$
$$= 2\left[\pi\delta(\omega) + \frac{1}{j\omega}\right] - 2\pi\delta(\omega) = \frac{2}{j\omega} \quad (3-42)$$

最后将常用信号的傅里叶变换列于表 3-1,以便查阅。

表 3-1 常用信号的傅里叶变换

序号	名称	时域表达式	傅里叶变换
1	矩形脉冲信号	$G_\tau(t) = E\left[u\left(t+\frac{\tau}{2}\right) - u\left(t-\frac{\tau}{2}\right)\right]$	$E\tau Sa\left(\frac{\omega\tau}{2}\right)$
2	单边指数信号	$e^{-at}u(t), a>0$	$\frac{1}{a+j\omega}$
3	双边指数信号	$e^{-a\|t\|}, a>0$	$\frac{2a}{a^2+\omega^2}$
4	单位冲激信号	$\delta(t)$	1
5	直流信号	1	$2\pi\delta(\omega)$
6	单位阶跃信号	$u(t)$	$\pi\delta(\omega) + \frac{1}{j\omega}$
7	符号函数	$\text{sgn}(t)$	$\frac{2}{j\omega}$
8	三角脉冲信号	$\begin{cases}(1-\frac{2\|t\|}{\tau}), & \|t\| < \frac{\tau}{2} \\ 0, & \|t\| > \frac{\tau}{2}\end{cases}$	$\frac{\tau}{2}Sa^2\left(\frac{\omega\tau}{2}\right)$
9	抽样脉冲信号	$Sa(\omega_0 t)$	$\begin{cases}\frac{\pi}{\omega_0}, & \|\omega\| < \omega_0 \\ 0, & \|\omega\| > \omega_0\end{cases}$

3.3 傅里叶变换的性质(Properties of Fourier Transform)

在信号分析的理论研究与实际设计工作中,经常需要了解当信号在时域进行某种运算后在频域会发生何种变化,或反过来,从频域的运算推测时域的变化。这时,可以根据式(3-29)来计算,也可以借助傅里叶变换的基本性质得出结果。后一种计算方法过程比较简便,而且物理概念清楚。因此,熟悉傅里叶变换的基本性质对于信号分析十分重要。本节将讨论连续时间傅里叶变换的基本性质,旨在通过这些性质揭示信号时域特性与频域特性之间的关系,同时掌握和运用这些性质可以简化傅里叶变换对的求取。

In this section, we consider a number of properties of the Fourier transform. These properties provide us with a significant amount of insight into the transform and into the relationship between the time-domain and frequency-domain descriptions of a signal. In addition, many of the properties are often useful in reducing the complexity of the evaluation of Fourier transforms or inverse transforms.

3.3.1 线性性质(Linearity)

If $f_1(t) \leftrightarrow F_1(\omega)$, $f_2(t) \leftrightarrow F_2(\omega)$, then

$$a_1 f_1(t) + a_2 f_2(t) \leftrightarrow a_1 F_1(\omega) + a_2 F_2(\omega) \tag{3-43}$$

where a_1, a_2 are constants.

Proof: According to the definition of the Fourier transform, this conclusion can be easily to prove.

$$\begin{aligned} \mathscr{F}[a_1 f_1(t) + a_2 f_2(t)] &= \int_{-\infty}^{\infty} [a_1 f_1(t) + a_2 f_2(t)] e^{-j\omega t} dt \\ &= \int_{-\infty}^{\infty} a_1 f_1(t) e^{-j\omega t} dt + \int_{-\infty}^{\infty} a_2 f_2(t) e^{-j\omega t} dt \\ &= a_1 F_1(\omega) + a_2 F_2(\omega) \end{aligned}$$

显然傅里叶变换是一种线性运算,它满足叠加性和齐次性。这个性质虽然简单,但很重要,它是频域分析的基础。在上节求符号函数的频谱时我们已经应用了此性质。式(3-43)可以推广到多个信号的线性组合。

Apparently, the Fourier transform is a linear operation, and it satisfies superposition and homogeneity. It is simple, but it is very important. It is the foundation of frequency domain analysis. The linearity property is easily extended to a linear combination of an arbitrary number of signals.

3.3.2 奇偶虚实性(Conjugation and Conjugate Symmetry)

The conjugation property states that if $f(t) \leftrightarrow F(\omega)$, then

$$f^*(t) \leftrightarrow F^*(-\omega) \tag{3-44}$$

Proof:

$$\begin{aligned} \mathscr{F}[f^*(t)] &= \int_{-\infty}^{\infty} f^*(t) e^{-j\omega t} dt \\ &= \left[\int_{-\infty}^{\infty} f(t) e^{j\omega t} dt \right]^* \\ &= F^*(-\omega) \end{aligned}$$

An important consequence of this property is that if the signal $f(t)$ is real, then $f^*(t) = f(t)$ and $F^*(-\omega) = F(\omega)$.

3.3.3 时移特性(Time Shifting)

If $f(t) \leftrightarrow F(\omega)$, then

$$f(t - t_0) \leftrightarrow e^{-j\omega t_0} F(\omega) \tag{3-45}$$

Proof: Replacing $t-t_0$ by x in Eq. (3-45), we obtain

$$\mathscr{F}[f(t-t_0)] = \int_{-\infty}^{\infty} f(x) e^{-j\omega(x+t_0)} dx$$

$$= e^{-j\omega t_0} \int_{-\infty}^{\infty} f(x) e^{-j\omega t} dx$$

$$= e^{-j\omega t_0} F(\omega)$$

similarly, we can prove $\mathscr{F}[f(t+t_0)] = e^{j\omega t_0} F(\omega)$.

从式(3-45)可以看出,在时域中信号右移 t_0,其频谱函数的幅度不变,而各频率分量的相位比原 $f(t)$ 各频率分量的相位滞后 ωt_0。

One consequence of the time-shift property is that a signal which is shifted in time does not have the magnitude of its Fourier transform altered. Thus, the effect of a time shift on a signal is to introduce into its transform a phase shift, namely, $-\omega t_0$, which is a linear function of ω.

Example 3-3: Determine the Fourier transform of the shifted impulse function $\delta(t-t_0)$.

Solution: Because $\delta(t) \leftrightarrow 1$, so that according to Eq. (3-45), we get

$$\delta(t-t_0) \leftrightarrow e^{-j\omega t_0}$$

Example 3-4: Determine the Fourier transform of three rectangular pulses shown in Fgure 3-14.

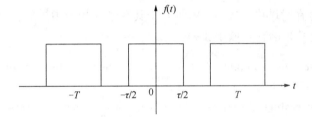

Figure 3-14 the waveform of three rectangular pulses

Solution: Let $g(t) = E\left[u\left(t+\dfrac{\tau}{2}\right) - u\left(t-\dfrac{\tau}{2}\right)\right]$, then

$$f(t) = g(t+T) + g(t) + g(t-T)$$

using the time shifting property, we get

$$\mathscr{F}[f(t)] = \mathscr{F}[g(t+T) + g(t) + g(t-T)]$$

$$= \left[E\tau Sa\left(\frac{\omega\tau}{2}\right)e^{j\omega T} + E\tau Sa\left(\frac{\omega\tau}{2}\right) + E\tau Sa\left(\frac{\omega\tau}{2}\right)e^{-j\omega T}\right]$$

$$= E\tau \cdot Sa\left(\frac{\omega\tau}{2}\right)[1 + e^{j\omega T} + e^{-j\omega T}]$$

$$= E\tau \cdot Sa\left(\frac{\omega\tau}{2}\right)[1 + 2\cos(\omega T)]$$

3.3.4 尺度特性(Time Scaling)

If $f(t) \leftrightarrow F(\omega)$, a is a constant($a \neq 0$), then

$$\mathscr{F}(at) \leftrightarrow \frac{1}{|a|} F\left(\frac{\omega}{a}\right) \tag{3-46}$$

Proof:
$$\mathscr{F}[f(at)] = \int_{-\infty}^{\infty} f(at) e^{-j\omega t} dt$$

We consider positive and negative values of a separately. For a is positive, and changing the variable of integration to $x = at$, we have

$$\mathscr{F}[f(at)] = \int_{-\infty}^{\infty} f(x) e^{-j\omega \frac{x}{a}} d\frac{x}{a}$$

$$= \frac{1}{a} \int_{-\infty}^{\infty} f(x) e^{-j\frac{\omega}{a} x} dx$$

$$= \frac{1}{a} F\left(\frac{\omega}{a}\right), \text{ for } a > 0$$

For a is negative, the upper and lower limits of the integral are opposite, thus

$$\mathscr{F}[f(at)] = -\frac{1}{a} F\left(\frac{\omega}{a}\right), \text{ for } a < 0$$

combining two results together, we obtain,

$$\mathscr{F}[f(at)] = \frac{1}{|a|} F\left(\frac{\omega}{a}\right)$$

especially, when $a = -1$, $\mathscr{F}[f(-t)] = F(-\omega)$.

由上可见，信号在时域中压缩($a>1$)等效于其频谱在频域中扩展；反之，信号在时域中扩展($a<1$)则等效于其频谱在频域中压缩。

Thus, aside from the amplitude factor $\frac{1}{|a|}$, a linear scaling in time by a factor of a corresponds to a linear scaling in frequency by a factor of $\frac{1}{a}$, and vice versa. Also, when $a = -1$, we see that reversing a signal in time also reverses its Fourier transform.

上述结论是不难理解的，因为信号的波形压缩 a 倍，信号随时间变化加快 a 倍，所以它所包含的频率分量增加 a 倍，也就是说频谱展宽 a 倍。根据能量守恒定理，各频率分量的大小必然减小 a 倍。这表明，信号的持续时间与其频带宽度成反比。图 3-15 列出了矩形脉冲的三种情况。

例 3-5：已知 $f(t) \leftrightarrow F(\omega)$，求 $f(2t+4)$ 的频谱函数。

解：信号 $f(2t+4)$ 是 $f(t)$ 经过压缩、平移两种基本运算而产生的信号，需要分别利用傅里叶变换的尺度特性和时移特性求其频谱。可以将 $f(t)$ 先进行压缩再平移，也可以将 $f(t)$ 先进行平移再压缩，两种方法的计算过程稍有不同，但结果一致。这里以前者为例，读者可自行练习后者。

先对 $f(t)$ 进行压缩 $t \to 2t$，利用尺度特性得

$$f(2t) \leftrightarrow \frac{1}{2} F\left(\frac{\omega}{2}\right)$$

再对 $f(2t)$ 进行左移 $t \to t+2$，利用时移特性得

$$f[2(t+2)] = f(2t+4) \leftrightarrow \frac{1}{2} F\left(\frac{\omega}{2}\right) e^{j2\omega}$$

从此例题分析可见，若 $f(t) \leftrightarrow F(\omega)$，则 $f(at+b)$ ($a \neq 0$) 的傅里叶变换为

$$\mathscr{F}[f(at+b)] = \frac{1}{|a|} F\left(\frac{\omega}{a}\right) e^{j\frac{b}{a}\omega} \qquad (3-47)$$

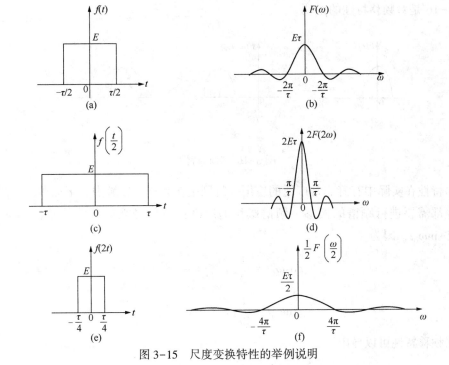

图 3-15 尺度变换特性的举例说明

3.3.5 频移特性(Frequency Shifting)

If $f(t) \leftrightarrow F(\omega)$, then
$$\mathscr{F}[f(t)e^{\pm j\omega_0 t}] = F(\omega \mp \omega_0) \qquad (3-48)$$

Proof: Consider $\mathscr{F}[f(t)e^{j\omega_0 t}]$,

$$\mathscr{F}[f(t)e^{j\omega_0 t}] = \int_{-\infty}^{\infty} f(t)e^{j\omega_0 t}e^{-j\omega t}dt$$
$$= \int_{-\infty}^{\infty} f(t)e^{-j(\omega-\omega_0)t}dt$$
$$= F(\omega - \omega_0)$$

similarly, $\mathscr{F}[f(t)e^{-j\omega_0 t}] = F(\omega+\omega_0)$. Where, ω_0 is a constant.

该性质表明信号在时域中与复因子 $e^{j\omega_0 t}$ 相乘,则在频域中将使这个频谱搬移 ω_0。也就是说,如果 $f(t)$ 的频谱原来在 $\omega=0$ 附近(低频信号),则将其乘以 $e^{j\omega_0 t}$ 就可以将其频谱搬移到 $\omega=\omega_0$ 附近,这个过程称为调制(modulation)。反之,如果 $f(t)$ 的频谱原来在 $\omega=\omega_0$ 附近(高频信号),则将 $f(t)$ 乘以 $e^{-j\omega_0 t}$ 就可以将其频谱搬移到 $\omega=0$ 附近,这个过程称为解调(demodulation)。如果 $f(t)$ 的频谱原来在 $\omega=\omega_1$ 附近,则乘以 $e^{j\omega_0 t}$ 后其频谱被搬移到 $\omega=\omega_0+\omega_1$ 附近,这样的过程称为变频(frequency conversion)。

The frequency ω_0 is referred to as the carrier frequency. The spectrum of the modulated output is simply that of the input, shifted in frequency by an amount equal to the carrier frequency ω_0. In frequency domain, this has the effect of shifting the spectrum of the modulated signal back to its original position on the frequency axis. The process of recovering the original signal from the modulated signal is referred to as demodulation.

图 3-16 是对频移特性的说明。

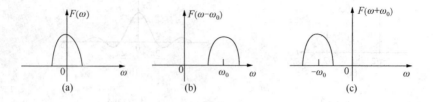

图 3-16 频移特性

频移特性在实际中有着非常广泛的应用，特别是在无线电领域中，诸如调制、混频、同步解调等都需要进行频谱的搬移。频谱搬移的原理是将信号乘以载波信号(carrier signal) $\cos\omega_0 t$ 或 $\sin\omega_0 t$，因为

$$\cos\omega_0 t = \frac{1}{2}[e^{j\omega_0 t} + e^{-j\omega_0 t}]$$

$$\sin\omega_0 t = \frac{1}{2j}[e^{j\omega_0 t} - e^{-j\omega_0 t}]$$

依据频移特性可以导出

$$f(t)\cos\omega_0 t \to \frac{1}{2}[F(\omega-\omega_0) + F(\omega+\omega_0)] \tag{3-49}$$

$$f(t)\sin\omega_0 t \to \frac{1}{2j}[F(\omega-\omega_0) - F(\omega+\omega_0)] \tag{3-50}$$

这两个式子表明了实际的调制过程中频谱搬移的情况。

In many situations, using a sinusoidal carrier of the form of $\cos\omega_0 t$ or $\sin\omega_0 t$ is often simpler than and equally as effective as using a complex exponential carrier.

Example 3-6: Find the Fourier transform of $e^{j\omega_0 t}$.

Solution: We have known that

$$1 \leftrightarrow 2\pi\delta(\omega)$$

using frequency shifting property, we can obtain

$$1 \cdot e^{j\omega_0 t} \leftrightarrow 2\pi\delta(\omega-\omega_0)$$

And from this result we can also deduce the Fourier transforms of cosine and sinusoidal signals, that is,

$$\cos\omega_0 t \leftrightarrow \pi[\delta(\omega-\omega_0) + \delta(\omega+\omega_0)] \tag{3-51}$$

$$\sin\omega_0 t \leftrightarrow j\pi[\delta(\omega+\omega_0) - \delta(\omega-\omega_0)] \tag{3-52}$$

3.3.6 对偶性(Duality)

If $\mathscr{F}[f(t)] = F(\omega)$, then

$$\mathscr{F}[F(t)] = 2\pi f(-\omega) \tag{3-53}$$

If $\mathscr{F}(t)$ is an even function, then

$$\mathscr{F}[F(t)] = 2\pi f(\omega).$$

Proof: Because
$$f(t) = \frac{1}{2\pi} \int_{-\infty}^{\infty} F(\omega) e^{j\omega t} d\omega$$
exchange the variables t and ω, that is, $t \leftrightarrow \omega$
$$2\pi f(\omega) = \int_{-\infty}^{\infty} F(t) e^{j\omega t} dt$$
then
$$2\pi f(-\omega) = \int_{-\infty}^{\infty} F(t) e^{-j\omega t} dt = \mathscr{F}[F(t)]$$
specifically, if $f(t)$ is an even function, then $\mathscr{F}[F(t)] = 2\pi f(\omega)$.

利用对偶特性，可以很方便地求某些信号的频谱，特别是有些直接利用定义无法求解的信号，往往可以利用对偶特性求得。

The duality property can also be used to determine or to suggest other properties of Fourier transforms. Specifically, if there are characteristics of a function of time that have implications with regard to the Fourier transform, then the same characteristics associated with a function of frequency will have dual implications in the time domain.

Example 3-7: Given the function $f(t) = \dfrac{1}{t}$, find its Fourier transform.

Solution: From previous section, we have known $\mathrm{sgn}(t) \leftrightarrow \dfrac{2}{j\omega}$. Then, according to the duality property, we obtain
$$\frac{2}{jt} \leftrightarrow 2\pi \mathrm{sgn}(-\omega)$$
thus,
$$\frac{1}{t} \leftrightarrow -j\pi \mathrm{sgn}(\omega)$$
that is,
$$\mathscr{F}\left[\frac{1}{t}\right] = -j\pi \mathrm{sgn}(\omega)$$

Example 3-8: Suppose that the Fourier transform of a function is
$$F(\omega) = E[u(\omega + \omega_0) - u(\omega - \omega_0)]$$
Determine the function $f(t)$.

Solution: Because $F(t) = E[u(t+\omega_0) - u(t-\omega_0)] \leftrightarrow 2E\omega_0 Sa(\omega\omega_0)$ according to the duality property, we have
$$F(t) \leftrightarrow 2\pi f(-\omega) = 2E\omega_0 Sa(\omega\omega_0)$$
thus,
$$f(-\omega) = \frac{E\omega_0}{\pi} Sa(\omega\omega_0) = f(\omega)$$
that is,
$$f(t) = f(\omega)|_{\omega=t} = \frac{E\omega_0}{\pi} Sa(\omega_0 t)$$

The two Fourier transform pairs and the relationship between them are depicted in Figure 3-17.

3.3.7 微分性质(Differentiation)

(1) 时域微分(differentiation in time domain)

If $\mathscr{F}[f(t)] = F(\omega)$, then

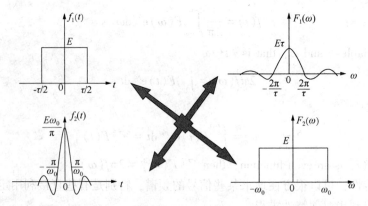

Figure 3-17　two Fourier transform pairs and the relationship between them

$$\mathscr{F}\left[\frac{\mathrm{d}}{\mathrm{d}t}f(t)\right] = j\omega F(\omega) \tag{3-54}$$

Proof: According to the inverse Fourier transform, we have

$$f(t) = \frac{1}{2\pi}\int_{-\infty}^{\infty} F(\omega) e^{j\omega t}\mathrm{d}\omega$$

$$\frac{\mathrm{d}}{\mathrm{d}t}f(t) = \frac{1}{2\pi}\int_{-\infty}^{\infty} F(\omega)\frac{\mathrm{d}}{\mathrm{d}t}e^{j\omega t}\mathrm{d}\omega$$

$$= \frac{1}{2\pi}\int_{-\infty}^{\infty} j\omega F(\omega) e^{j\omega t}\mathrm{d}\omega$$

this implies

$$\mathscr{F}\left[\frac{\mathrm{d}}{\mathrm{d}t}f(t)\right] = j\omega F(\omega)$$

We can extend this result to nth order derivatives by repeated differentiations within the integral. Thus

$$\frac{\mathrm{d}^n f(t)}{\mathrm{d}t^n} \leftrightarrow (j\omega)^n F(\omega)$$

此性质表明，在时域对信号 $f(t)$ 求导数，对应于频域中用 $j\omega$ 乘 $f(t)$ 的频谱函数。如果应用此性质对微分方程两端求变换，即可将微分方程变换成代数方程，简化求解过程。

Example 3-9: Suppose that we wish to calculate the Fourier transform $F(\omega)$ for the signal $f(t)$ displayed in Figure 3-18 rather than applying the Fourier integral directly to $f(t)$, we instead consider the signal $\frac{\mathrm{d}}{\mathrm{d}t}f(t)$.

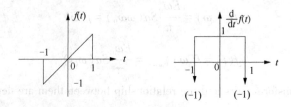

Figure 3-18　the waveform of Example 3-9

As illustrated in Figure 3-18, $\frac{\mathrm{d}}{\mathrm{d}t}f(t)$ is the sum of a rectangular pulse and two impulses.

$$\frac{\mathrm{d}f(t)}{\mathrm{d}t} = G_2(t) - \delta(t+1) - \delta(t-1)$$

taking the Fourier transform on both sides, we get

$$j\omega F(\omega) = 2Sa(\omega) - e^{j\omega} - e^{-j\omega}$$
$$= 2Sa(\omega) - 2\cos(\omega)$$

so that

$$F(\omega) = \frac{2}{j\omega}[Sa(\omega) - \cos\omega]$$

利用时域微分特性还容易求出一些在通常意义下不易求得的变换关系，例如：
由 $\delta(t) \leftrightarrow 1$，可得

$$\delta'(t) \leftrightarrow j\omega \tag{3-55}$$

$$\delta^{(n)}(t) \leftrightarrow (j\omega)^n \tag{3-56}$$

由 $\frac{1}{t} \leftrightarrow -\pi j\mathrm{sgn}(\omega)$，可得

$$\frac{1}{t^2} \leftrightarrow -\pi\omega\mathrm{sgn}(\omega) = -\pi|\omega| \tag{3-57}$$

(2) 频域微分 (differentiation in frequency domain)

If $\mathscr{F}[f(t)] = F(\omega)$, then

$$-jtf(t) \leftrightarrow \frac{\mathrm{d}F(\omega)}{\mathrm{d}\omega} \quad \text{or} \quad tf(t) \leftrightarrow j\frac{\mathrm{d}F(\omega)}{\mathrm{d}\omega} \tag{3-58}$$

Proof: Beginning with the Fourier transform

$$F(\omega) = \int_{-\infty}^{\infty} f(t)\mathrm{e}^{-j\omega t}\mathrm{d}t$$

we differentiate both sides of this equation with respect to ω and obtain

$$\frac{\mathrm{d}}{\mathrm{d}\omega}F(\omega) = \int_{-\infty}^{\infty} -jtf(t)\mathrm{e}^{-j\omega t}\mathrm{d}t$$

from which it follows that

$$-jtf(t) \leftrightarrow \frac{\mathrm{d}F(\omega)}{\mathrm{d}\omega} \quad \text{or} \quad tf(t) \leftrightarrow j\frac{\mathrm{d}F(\omega)}{\mathrm{d}\omega}$$

similarly,

$$(-jt)^n f(t) \leftrightarrow \frac{\mathrm{d}^n F(\omega)}{\mathrm{d}\omega^n} \quad \text{or} \quad t^n f(t) \leftrightarrow j^n \frac{\mathrm{d}^n F(\omega)}{\mathrm{d}\omega^n}$$

Example 3-10: Use the frequency differentiation property to find the Fourier transform of $f(t) = t\mathrm{e}^{-\alpha t}u(t)$.

Solution: Because

$$\mathrm{e}^{-\alpha t}u(t) \leftrightarrow F(\omega) = \frac{1}{\alpha + j\omega}$$

therefore,

$$t\mathrm{e}^{-\alpha t}u(t) \leftrightarrow j\frac{\mathrm{d}F(\omega)}{\mathrm{d}\omega} = j\left(\frac{1}{\alpha + j\omega}\right)' = \frac{1}{(\alpha + j\omega)^2}$$

3.3.8 积分性质 (Integration)

If $f(t) \leftrightarrow F(\omega)$, then

$$\int_{-\infty}^{t} f(\tau)\mathrm{d}\tau \leftrightarrow \left[\pi\delta(\omega) + \frac{1}{j\omega}\right] \cdot F(\omega) = \pi F(0)\delta(\omega) + \frac{F(\omega)}{j\omega} \quad (3-59)$$

Proof:
$$\mathscr{F}\left[\int_{-\infty}^{t} f(\tau)\mathrm{d}\tau\right] = \int_{-\infty}^{\infty}\left[\int_{-\infty}^{t} f(\tau)\mathrm{d}\tau\right]e^{-j\omega t}\mathrm{d}t$$

$$= \int_{-\infty}^{\infty}[f(t) * u(t)]e^{-j\omega t}\mathrm{d}t$$

$$= \int_{-\infty}^{\infty}\left[\int_{-\infty}^{\infty} f(\tau)u(t-\tau)\mathrm{d}\tau\right]e^{-j\omega t}\mathrm{d}t$$

exchanging the order of the integral, we obtain

$$\mathscr{F}\left[\int_{-\infty}^{t} f(\tau)\mathrm{d}\tau\right] = \int_{-\infty}^{\infty} f(\tau)\left[\int_{-\infty}^{\infty} u(t-\tau)e^{-j\omega t}\mathrm{d}t\right]\mathrm{d}\tau$$

$$= \int_{-\infty}^{\infty} f(\tau)\left[\pi\delta(\omega) + \frac{1}{j\omega}\right]e^{-j\omega\tau}\mathrm{d}\tau$$

$$= \left[\pi\delta(\omega) + \frac{1}{j\omega}\right]F(\omega)$$

$$= \pi F(0)\delta(\omega) + \frac{F(\omega)}{j\omega}$$

时域积分性质多用于 $F(0)=0$ 的情况，$F(0)=0$ 表明 $f(t)$ 的频谱函数中直流分量的频谱分量为零。由于 $F(\omega) = \int_{-\infty}^{\infty} f(t)e^{-j\omega t}\mathrm{d}t$，显然有 $F(0) = \int_{-\infty}^{\infty} f(t)\mathrm{d}t$。也就是说，$F(0) = 0$ 等效于 $\int_{-\infty}^{\infty} f(t)\mathrm{d}t = 0$，即当 $f(t)$ 波形在 t 轴上下两部分面积相等时，$F(0)=0$，从而有

$$\int_{-\infty}^{t} f(\tau)\mathrm{d}\tau \leftrightarrow \frac{F(\omega)}{j\omega}$$

Example 3-11: Calculate the Fourier transform of the unit step signal using the integration property.

Solution: Because $u(t) = \int_{-\infty}^{t}\delta(t)\mathrm{d}t$ and $\delta(t)\leftrightarrow 1$, then

$$u(t) \leftrightarrow \left[\pi\delta(\omega) + \frac{1}{j\omega}\right] \cdot 1 = \pi\delta(\omega) + \frac{1}{j\omega}$$

3.3.9 卷积定理(Convolution Theorem)

(1) 时域卷积(time convolution)

If $f_1(t)\leftrightarrow F_1(\omega)$, $f_2(t)\leftrightarrow F_2(\omega)$, then

$$f_1(t) * f_2(t) \leftrightarrow F_1(\omega) \cdot F_2(\omega) \quad (3-60)$$

Proof: According to the definition of the convolution, we have

$$f_1(t) * f_2(t) = \int_{-\infty}^{\infty} f_1(\tau)f_2(t-\tau)\mathrm{d}\tau$$

then

$$\mathscr{F}[f_1(t) * f_2(t)] = \int_{-\infty}^{\infty}\left[\int_{-\infty}^{\infty} f_1(\tau)f_2(t-\tau)\mathrm{d}\tau\right]e^{-j\omega t}\mathrm{d}t$$

exchanging the order of the integral, we obtain

$$\mathscr{F}[f_1(t)*f_2(t)] = \int_{-\infty}^{\infty} f_1(\tau) \left[\int_{-\infty}^{\infty} f_2(t-\tau)e^{-j\omega t}dt\right]d\tau$$

from the time shifting property, we get

$$\mathscr{F}[f_1(t)*f_2(t)] = \int_{-\infty}^{\infty} f_1(\tau)F_2(\omega)e^{-j\omega\tau}d\tau$$

therefore

$$\mathscr{F}[f_1(t)*f_2(t)] = F_1(\omega)F_2(\omega)$$

式(3-60)称为时域卷积定理,它说明两个时间函数卷积的频谱等于各个时间函数频谱的乘积,即在时域中两信号的卷积等效于在频域中频谱相乘。时域卷积定理是傅里叶变换中最重要的性质之一,在分析线性时不变系统中有着重要的意义,它是滤波技术的理论基础。

Eq. (3-60) is of major importance in signal and system analysis. As expressed in this equation, the Fourier transform maps the convolution of two signals into the product of their Fourier transforms. Time domain convolution theorem is the theoretical basis for filtering technique.

(2) 频域卷积(frequency convolution)

If $f_1(t) \leftrightarrow F_1(\omega)$, $f_2(t) \leftrightarrow F_2(\omega)$, then

$$f_1(t) \cdot f_2(t) \leftrightarrow \frac{1}{2\pi}F_1(\omega) * F_2(\omega) \tag{3-61}$$

Proof:

$$\mathscr{F}[f_1(t) \cdot f_2(t)] = \int_{-\infty}^{\infty} f_1(t)f_2(t)e^{-j\omega t}dt$$

using the inverse Fourier transform to express $f_1(t)$, then we get

$$\mathscr{F}[f_1(t)f_2(t)] = \int_{-\infty}^{\infty}\left[\frac{1}{2\pi}\int_{-\infty}^{\infty}F_1(u)e^{jut}du\right]f_2(t)e^{-j\omega t}dt$$

exchanging the order of the integral, we obtain

$$\mathscr{F}[f_1(t)f_2(t)] = \frac{1}{2\pi}\int_{-\infty}^{\infty}F_1(u)\left[\int_{-\infty}^{\infty}f_2(t)e^{-j(\omega-u)t}dt\right]du$$

$$= \frac{1}{2\pi}\int_{-\infty}^{\infty}F_1(u)F_2(\omega-u)du$$

$$= \frac{1}{2\pi}F_1(\omega) * F_2(\omega)$$

式(3-61)称为频域卷积定理,它说明两个时间信号乘积的频谱等于两个信号的频谱函数的卷积乘以$\frac{1}{2\pi}$。显然时域与频域卷积定理是对称的,这是由傅里叶变换的对偶性所决定的。

Because of duality between the time and frequency domains, we would expect a dual property also to hold as expressed in Eq. (3-61) (i.e., that multiplication in the time domain corresponds to convolution in the frequency domain).

最后,将上面讨论的傅里叶变换的性质列于表3-2中,以便查阅。

表 3-2　傅里叶变换性质

序号	性质名称	时域	频域
1	线性性质	$a_1 f_1(t) + a_2 f_2(t)$	$a_1 F_1(\omega) + a_2 F_2(\omega)$
2	奇偶虚实性	$f^*(t)$	$F^*(-\omega)$
3	尺度特性	$f(at)$，$a \neq 0$	$\dfrac{1}{\|a\|} F\left(\dfrac{\omega}{a}\right)$
4	时移特性	$f(t-t_0)$	$e^{-j\omega t_0} F(\omega)$
5	时移+尺度特性	$f(at+b)$，$a \neq 0$	$\dfrac{1}{\|a\|} F\left(\dfrac{\omega}{a}\right) e^{j\frac{b}{a}\omega}$
6	频移特性	$f(t) e^{\pm j\omega_0 t}$	$F(\omega \mp \omega_0)$
7	对偶性	$F(t)$	$2\pi f(-\omega)$
8	时域微分性质	$\dfrac{d^n f(t)}{dt^n}$	$(j\omega)^n F(\omega)$
9	频域微分性质	$t^n f(t)$	$j^n \dfrac{d^n F(\omega)}{d\omega^n}$
10	积分性质	$\int_{-\infty}^{t} f(\tau) d\tau$	$\pi F(0) \delta(\omega) + \dfrac{F(\omega)}{j\omega}$
11	时域卷积定理	$f_1(t) * f_2(t)$	$F_1(\omega) F_2(\omega)$
12	频域卷积定理	$f_1(t) \cdot f_2(t)$	$\dfrac{1}{2\pi} F_1(\omega) * F_2(\omega)$

3.4　周期信号的傅里叶变换（Fourier Transform of Periodic Signals）

在频域分析中，如果对周期信号用傅里叶级数表示，对非周期信号用傅里叶变换表示，显然会给频域分析带来很多不便。那么，能否统一起来？这就需要讨论周期信号是否存在傅里叶变换。虽然周期信号不满足绝对可积条件，但和之前讨论的直流信号、阶跃信号等一样，在引入奇异信号之后，是存在傅里叶变换的。事实上，周期信号的傅里叶变换可以从它的指数形式的傅里叶级数得到，是由一系列的冲激函数组成，其冲激强度正比于傅里叶级数的系数。

In the preceding section, we introduced the Fourier transform representation and gave several examples. While our attention in that section was focused on aperiodic signals, we can also develop Fourier transform representation for periodic signals, thus allowing us to consider both periodic and aperiodic signals within a unified context. In fact, as we will see, we can construct the Fourier transform of a periodic signal directly from its Fourier series representation. The resulting transform consists of a train of impulses in the frequency domain, with the areas of the impulses proportional to the Fourier series coefficients.

设 $f(t)$ 为周期信号，其周期为 T，且 $T = \dfrac{2\pi}{\omega_0}$，根据周期信号的指数形式傅里叶级数可将其表示为

$$f(t) = \sum_{n=-\infty}^{\infty} F_n e^{jn\omega_0 t} \tag{3-62}$$

其中，ω_0 为基波角频率，F_n 为复振幅，且

$$F_n = \frac{1}{T}\int_{-\frac{T}{2}}^{\frac{T}{2}} f(t)\mathrm{e}^{-jn\omega_0 t}\mathrm{d}t \tag{3-63}$$

对周期信号 $f(t)$ 进行傅里叶变换，从而有

$$F(\omega) = \mathscr{F}[f(t)] = \mathscr{F}\Big[\sum_{n=-\infty}^{\infty} F_n \mathrm{e}^{jn\omega_0 t}\Big]$$

$$= \sum_{n=-\infty}^{\infty} F_n \mathscr{F}[\mathrm{e}^{jn\omega_0 t}]$$

根据傅里叶变换的频移特性，可知

$$\mathrm{e}^{jn\omega_0 t} \leftrightarrow 2\pi\delta(\omega - n\omega_0)$$

所以

$$F(\omega) = 2\pi \sum_{n=-\infty}^{\infty} F_n \delta(\omega - n\omega_0)$$

上式表明，周期信号的频谱由无限多个冲激函数组成，各冲激函数位于周期信号 $f(t)$ 的各次谐波 $n\omega_0$ 处，且冲激强度为 $|F(\omega)|$ 的 2π 倍。显然，周期信号的频谱是离散的，这与之前的结论一致。

下面分析非周期信号的傅里叶变换与周期信号的傅里叶级数系数之间的关系。

设 $f_0(t)$ 是从周期信号 $f(t)$ 中提取出的一个周期，则 $f_0(t)$ 是非周期信号，其傅里叶变换可由定义求得

$$F_0(\omega) = \int_{-\frac{T}{2}}^{\frac{T}{2}} f_0(t)\mathrm{e}^{-j\omega t}\mathrm{d}t \tag{3-64}$$

比较式(3-63)和式(3-64)，在 $\left(-\dfrac{T}{2}, \dfrac{T}{2}\right)$ 内，$f_0(t)$ 和 $f(t)$ 相同，有

$$F_n = \frac{1}{T}F_0(\omega)\bigg|_{\omega = n\omega_0} \tag{3-65}$$

由此看出，周期信号的傅里叶级数的系数 F_n 等于单个周期信号的傅里叶变换 $F_0(\omega)$ 在 $n\omega_0$ 频率点的值乘以 $\dfrac{1}{T}$。所以可利用单个周期信号的傅里叶变换方便求出周期性信号的傅里叶级数的系数。

As a result, the Fourier series coefficient F_n of periodic signal is equal to the value in frequency $n\omega_0$ of the Fourier transform of the signal in a period multiplied by $\dfrac{1}{T}$. So the Fourier series coefficient of periodic signal can be easily obtained from the Fourier transform of the signal in a period.

Example 3-12: A special kind of periodic signal is the periodic unit impulse signal illustrated in Figure 3-19(a). Find the Fourier transform of it.

Solution: Because $\delta(t) \leftrightarrow 1$, so the Fourier series coefficient of $\delta_T(t)$ is

$$F_n = \frac{1}{T}\mathscr{F}[\delta(t)]\Big|_{\omega = n\omega_0} = \frac{1}{T}$$

expanding $\delta_T(t)$ to Fourier series, we get

$$\delta_T(t) = \sum_{n=-\infty}^{\infty} F_n \mathrm{e}^{jn\omega_0 t} = \frac{1}{T}\sum_{n=-\infty}^{\infty} \mathrm{e}^{jn\omega_0 t}$$

thus

$$F(\omega) = \mathscr{F}[\delta_T(t)] = \frac{1}{T}\sum_{n=-\infty}^{\infty} 2\pi\delta(\omega - n\omega_0)$$

$$= \frac{2\pi}{T}\sum_{n=-\infty}^{\infty} \delta(\omega - n\omega_0)$$

$$= \omega_0 \sum_{n=-\infty}^{\infty} \delta(\omega - n\omega_0)$$

The Fourier transform of $\delta_T(t)$ is depicted in Figure 3-19(b).

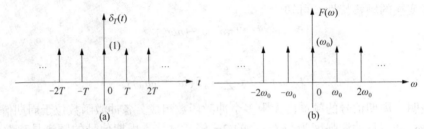

Figure 3-19 the periodic unit impulse signal and its Fourier transform

Example 3-13: Find the Fourier transform of a periodic pulse signal as shown in Figure 3-20.

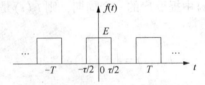

Figure 3-20 the periodic pulse signal

Solution: The Fourier series representation for this function has been found previously as

$$f(t) = \sum_{n=-\infty}^{\infty} \frac{E\tau}{T} Sa\left(\frac{n\omega_0\tau}{2}\right) e^{jn\omega_0 t}$$

and the Fourier series coefficient

$$F_n = \frac{E\tau}{T} Sa\left(\frac{n\omega_0\tau}{2}\right)$$

therefore, the Fourier transform of $f(t)$ is

$$\mathscr{F}[f(t)] = \frac{2\pi E\tau}{T}\sum_{n=-\infty}^{\infty} Sa\left(\frac{n\omega_0\tau}{2}\right)\delta(\omega - n\omega_0)$$

$$= E\tau\omega_0 \sum_{n=-\infty}^{\infty} Sa\left(\frac{n\omega_0\tau}{2}\right)\delta(\omega - n\omega_0)$$

3.5 连续信号的抽样定理（The Sampling Theorem for Continuous Signal）

在一定条件下，一个连续时间信号完全可以由它在等间隔时刻上的抽样值重建。这一点抽样定理给出了严格的描述。这个定理极其重要，当今电影、电视都建立在该定理的基础之上。其次，它建立了连续时间信号和离散时间信号之间联系的桥梁。在数字计算机和数字系统飞速发展的今天，物理世界的任何模拟量必须首先变为数字信号，才能有效地在数字计算机中处理，因此，掌握抽样理论是非常有意义的。

Under certain conditions, a continuous-time signal can be completely represented by and recoverable from knowledge of its values, or samples, at points equally spaced in time. This somewhat surprising property follows from a basic result that is referred to as the sampling theorem. This theorem is extremely important and useful. Today, movies and TV are based on the sampling theorem. Much of the importance of the sampling theorem also lies in its role as a bridge between continuous-time signals and discrete-time signals. Therefore, mastering the sampling theorem is very important.

3.5.1 时域抽样定理(The Sampling Theorem in Time Domain)

时域抽样就是利用抽样脉冲(sampling pulse)信号 $p(t)$ 从时域连续信号 $f(t)$ 中抽取一系列的离散样值，即抽样信号 $f_s(t)$。信号的抽样可通过抽样器(sampling device)来实现。抽样器本质上是一开关，如图 3-21 所示，开关每隔 T_s 接通输入信号，接通的时间是 τ。显然，抽样器输出的信号 $f_s(t)$ 只包含开关接通时间内输入信号 $f(t)$ 的一些小段，这些小段就是原输入信号的取样。

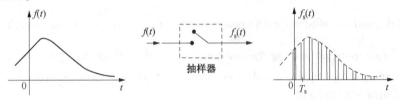

图 3-21 信号的抽样

抽样过程实际是相乘过程，可用连续信号 $f(t)$ 与开关函数 $p(t)$ 相乘来表示，抽样以后的信号表达式为

$$f_s(t) = f(t) \cdot p(t) \tag{3-66}$$

下面讨论矩形脉冲信号抽样。

(1) 矩形脉冲信号抽样(rectangular pulse signal sampling)

如果抽样脉冲信号是周期为 T_s，幅度为 1，宽度为 τ 的矩形脉冲信号 $p(t)$，由上一节可知其傅里叶变换为

$$P(\omega) = \frac{2\pi\tau}{T_s} \sum_{n=-\infty}^{\infty} Sa\left(\frac{n\omega_s \tau}{2}\right) \delta(\omega - n\omega_s) \tag{3-67}$$

其中 $\omega_s = \frac{2\pi}{T_s}$。

设原输入信号 $f(t)$ 的频谱密度函数为 $F(\omega)$，则根据频域卷积定理可得抽样信号 $f_s(t)$ 的傅里叶变换

$$\begin{aligned} F_s(\omega) &= \frac{1}{2\pi} F(\omega) * P(\omega) \\ &= \frac{1}{2\pi} \cdot F(\omega) * \frac{2\pi\tau}{T_s} \sum_{n=-\infty}^{\infty} Sa\left(\frac{n\omega_s \tau}{2}\right) \delta(\omega - n\omega_s) \\ &= \frac{\tau}{T_s} \sum_{n=-\infty}^{\infty} Sa\left(\frac{n\omega_s \tau}{2}\right) F(\omega - n\omega_s) \end{aligned} \tag{3-68}$$

由此可见,抽样信号的频谱由原信号频谱的无限个频移所组成,其频移的角频率为 $n\omega_s$, $n=0$, ±1, ±2, $\cdots$,它们的振幅随着 $\dfrac{\tau}{T_s}Sa\left(\dfrac{n\omega_s\tau}{2}\right)$ 的变化而变化。

(2) 冲激信号抽样(impulse train sampling)

在抽样脉冲信号 $p(t)$ 中,当脉冲宽度 τ 很小时,抽样脉冲信号可以近似看成是周期单位冲激信号,通常把这种抽样称为冲激抽样或理想抽样(ideal sampling)。

设周期单位冲激信号为

$$\delta_{T_s}(t) = \sum_{n=-\infty}^{\infty} \delta(t-nT_s) \qquad (3-69)$$

则抽样信号为

$$f_s(t) = f(t) \cdot \delta_{T_s}(t) = f(t) \cdot \sum_{n=-\infty}^{\infty} \delta(t-nT_s)$$

$$= \sum_{n=-\infty}^{\infty} f(nT_s)\delta(t-nT_s) \qquad (3-70)$$

可以看出,抽样信号 $f_s(t)$ 是原信号 $f(t)$ 在 $t=0$, $\pm T_s$, $\pm 2T_s$, $\cdots$ 处的一些离散值。

The periodic impulse train $\delta_{T_s}(t)$ is referred to as the sampling function, the period T_s as the sampling period, and the fundamental frequency of $\delta_{T_s}(t)$, $\omega_s = \dfrac{2\pi}{T_s}$, as the sampling frequency. According to the sampling property of the unit impulse signal, we see that $f_s(t)$ is an impulse train with the amplitudes of the impulse equal to the samples of $f(t)$ at intervals spaced by T_s.

由上一节可知,$\delta_{T_s}(t)$ 的频谱为

$$\delta_{T_s}(\omega) = \dfrac{2\pi}{T_s}\sum_{n=-\infty}^{\infty}\delta(\omega-n\omega_s)$$

所以抽样信号的频谱为

$$F_s(\omega) = \dfrac{1}{T_s}F(\omega) * \sum_{n=-\infty}^{\infty}\delta(\omega-n\omega_s) = \dfrac{1}{T_s}\sum_{n=-\infty}^{\infty}F(\omega-n\omega_s) \qquad (3-71)$$

上式说明,$F_s(\omega)$ 是频率的周期函数,是由原信号频谱的无限个频移组成的,如图 3-22 所示。

That is, $F_s(\omega)$ is a periodic function of ω consisting of a superposition of shifted replicas of $F(\omega)$, scaled by $\dfrac{1}{T_s}$, as illustrated in Figure 3-22.

(3) 时域抽样定理(the sampling theorem in time domain)

连续信号被抽样以后,抽样信号是否保留了原信号 $f(t)$ 的全部信息,在什么条件下才能保留原信号的全部信息,时域抽样定理给出了答案,下面介绍该定理。

时域抽样定理:一个频谱受限的信号 $f(t)$,如果其频谱只占据 $-\omega_m$ 到 ω_m 的范围,则信号 $f(t)$ 可以用等间隔的抽样值唯一地表示,而抽样间隔 $T_s \leq \dfrac{1}{2f_m}$(其中 $\omega_m=2\pi f_m$),或者说,最低抽样频率为 $2f_m$。

Let $f(t)$ be a band-limited signal with $F(\omega)=0$ for $|\omega|>\omega_m$. Then $f(t)$ is uniquely determined by its samples $f(nT_s)$, $n=0$, ±1, ±2, $\cdots$, if the sampling interval

$$T_s \leqslant \frac{1}{2f_m} \text{ (or } \omega_s \geqslant 2\omega_m, f_s \geqslant 2f_m)$$

where,

$$\omega_m = 2\pi f_m, \quad \omega_s = \frac{2\pi}{T_s}$$

由时域抽样定理及图 3-22 可知，在 $\omega_s \geqslant 2\omega_m$ 的条件下，周期性频谱无混叠现象。将抽样信号的频谱通过一个截止频率大于 ω_m，但小于 $\omega_s - \omega_m$，且增益为 T_s 的理想低通滤波器，就能完全地将 $f(t)$ 恢复出来。

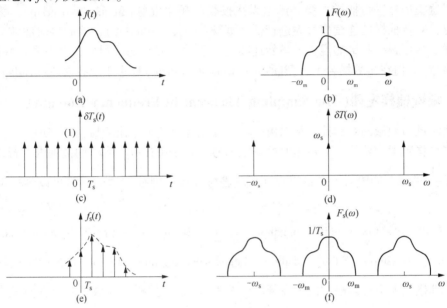

图 3-22 理想抽样

Given these samples, we can reconstruct $f(t)$ by generating a periodic impulse train in which successive impulse have amplitudes that are successive sample values. This impulse train is then processes through an ideal lowpass filter with gain T_s and cutoff frequency greater than ω_m and less than $\omega_s - \omega_m$. The resulting output signal will exactly equal $f(t)$.

因此，能从 $f_s(t)$ 中恢复 $f(t)$ 的唯一条件是

$$T_s \leqslant \frac{1}{2f_m} \text{ (或 } \omega_s \geqslant 2\omega_m, f_s \geqslant 2f_m) \tag{3-72}$$

抽样的最大允许间隔 $T_s = \frac{1}{2f_m}$ 称为"奈奎斯特间隔"(Nyquist interval)，或者说抽样的最低允许频率 $f_s = 2f_m$，称为"奈奎斯特频率"(Nyquist frequency)。当抽样角频率 $\omega_s > 2\omega_m$ 时，称为过采样 (over sampling)，$\omega_s < 2\omega_m$ 时称为欠采样 (undersampling)，$\omega_s = 2\omega_m$ 时为临界采样 (critical sampling)。

例 3-14：已知实信号 $f(t)$ 的最高频率为 f_m，试分别计算对下列信号抽样时，不发生混叠的最小抽样频率 f_{smin}：

① $f(2t)$ ② $f(t) * f(2t)$ ③ $f(t) \cdot f(2t)$ ④ $f(t) + f(2t)$

解：信号在时域的压缩对应其频谱在频域的扩展，故信号 $f(2t)$ 的最高频率为 $2f_m$。

① 根据抽样定理，对信号 $f(2t)$ 抽样时，最小抽样频率 $f_{smin} = 4f_m$。

② 信号在时域的卷积对应其频谱的乘积，故信号 $f(t) * f(2t)$ 的最高频率为 f_m，对信号 $f(t) * f(2t)$ 抽样时，最小抽样频率 $f_{smin} = 2f_m$。

③ 信号在时域的乘积对应其频谱的卷积，故信号 $f(t) \cdot f(2t)$ 的最高频率为 $3f_m$，对信号 $f(t) \cdot f(2t)$ 抽样时，最小抽样频率 $f_{smin} = 6f_m$。

④ 信号在时域的相加对应其频谱的相加，故信号 $f(t) + f(2t)$ 的最高频率为 $2f_m$，对信号 $f(t) + f(2t)$ 抽样时，最小抽样频率 $f_{smin} = 4f_m$。

在实际工程中，许多信号的频谱很宽或无限宽。如果在不满足抽样定理约束条件的情况下直接对这类信号进行抽样，势必产生无法接受的频率混叠（frequency overlap）。为了改善这种情况，对待抽样的连续信号先进行低通滤波（lowpass filtering），然后再对滤波后的信号进行抽样，从而减少频谱的混叠。虽然连续信号经过抗混叠滤波器后，会损失一些信息（称为截断误差）。但在多数场合下，截断误差（truncation error）远小于混叠误差（overlap error）。

3.5.2 频域抽样定理（The Sampling Theorem in Frequency Domain）

根据时域与频域的对称性，可以由时域抽样定理直接推论出频域抽样定理。

频域抽样定理：若信号 $f(t)$ 是时间受限信号，它集中在 $-t_m$ 到 t_m 的范围内，若在频域中以不大于 $\frac{1}{2t_m}$ 的频率间隔对 $f(t)$ 的频谱 $F(\omega)$ 进行抽样，则抽样后的频谱可以唯一地表示原信号。

Let $f(t)$ be a time-limited signal with $f(t) = 0$ for $|t| > t_m$. Then $F(\omega)$ is uniquely determined by its samples $F_s(\omega)$, if the frequency sampling interval is no greater than $\frac{1}{2t_m}$.

从物理概念上不难理解，因为在频域中对 $F(\omega)$ 进行抽样，等效于 $f(t)$ 在时域中重复形成周期信号。只要抽样间隔不大于 $\frac{1}{2t_m}$，则在时域中波形就不会产生混叠。用矩形脉冲作为选通信号从周期信号中选出单个脉冲就可以无失真地恢复出原信号了。

信号的时域抽样定理和频域抽样定理是信号处理的重要内容，其从理论上阐述了信号的时域与频域之间的内在联系，奠定了利用数字化方法分析信号与系统的理论基础，在信号与系统的广泛应用中发挥了重要作用。

Time domain and frequency domain sampling theoremare the important parts of signal processing, which describe the intrinsic relationship of signals in time domain and frequency domain, establish the theoretical foundation by using digital method for the analysis of signal and system, play an important role in the wide applications of signals and systems.

3.6 LTI 系统的频域分析（Fourier Analysis of Continuous-time LTI Systems）

在第 2 章我们已经介绍了系统的时域分析方法，时域分析是在时域中求解系统的响应，它反映了输入信号 $e(t)$ 通过系统后，其输出信号 $r(t)$ 随时间变化的规律。本节将以信号的频谱分析为基础，讨论系统的频域分析方法。频域分析是在频域中求解系统的响应，它反映了输入信号的频谱通过系统后，输出信号频谱随频率变化的情况。

3.6.1 系统的频率响应(Frequency Response of the System)

从系统的时域分析可知，对于一个线性时不变系统，如果外加激励为 $e(t)$，零状态响应为 $r_{zs}(t)$，则 $r_{zs}(t)$ 等于 $e(t)$ 与系统的单位冲激响应 $h(t)$ 的卷积，即

$$r_{zs}(t) = e(t) * h(t) \tag{3-73}$$

若令

$$\begin{cases} E(\omega) \leftrightarrow e(t) \\ H(\omega) \leftrightarrow h(t) \\ R_{zs}(\omega) \leftrightarrow r_{zs}(t) \end{cases}$$

根据卷积定理得

$$R_{zs}(\omega) = E(\omega) \cdot H(\omega) \tag{3-74}$$

式中 $H(\omega)$ 为该系统单位冲激响应 $h(t)$ 的傅里叶变换，系统单位冲激响应 $h(t)$ 表征的是系统时域特性，而 $H(\omega)$ 表征的是系统频域特性。所以 $H(\omega)$ 称作系统的频率响应(frequency response)，也称系统函数(system function)。由式(3-74)可以得出 $H(\omega)$ 的定义

$$H(\omega) = \frac{R_{zs}(\omega)}{E(\omega)} = |H(\omega)| e^{j\varphi(\omega)} \tag{3-75}$$

即系统函数是系统零状态响应的傅里叶变换与激励信号傅里叶变换之比。式中，$|H(\omega)|$ 是系统的幅频特性，$\varphi(\omega)$ 是系统的相频特性，统称为系统的频率特性。

In general, the system function $H(\omega)$ is a complex function and can be written in polar form like Eq. (3-75). Where $|H(\omega)|$ is known as amplitude-frequency response and $\varphi(\omega)$ is known as phase-frequency response.

We can also get

$$|R(\omega)| = |E(\omega)| \cdot |H(\omega)|$$
$$\varphi_r(\omega) = \varphi_e(\omega) + \varphi_h(\omega)$$

These two equations imply that output spectrums, both amplitude-frequency and phase-frequency, have been modified by the system function of the system. The output amplitude spectrum is weighted by $|H(\omega)|$, and the output phase spectrum is weighted by $\varphi_h(\omega)$.

系统的频率特性与实信号的频谱密度函数的特性相类似。但也有不同之处，系统带宽(bandwith)(不同于信号带宽)一般定义为等于 $|H(\omega)|$ 最大值的 $\frac{1}{\sqrt{2}}$ 处的频率为 ω_c [称为半功率点频率(half power frequency)，或截止频率(cutoff frequency)或3dB频率(3dB frequency)]作为系统带宽的依据。由系统不同的表示形式，可以用不同的方法得到系统函数。

(1) 用微分方程表征的系统(the differential equation representation of system)

工程实际中有相当广泛的线性时不变系统其输入输出关系可以由一个线性常系数微分方程表述

$$\sum_{k=0}^{n} a_k \frac{d^k r(t)}{dt^k} = \sum_{k=0}^{m} b_k \frac{d^k e(t)}{dt^k} \tag{3-76}$$

对上式两端同时求傅里叶变换，由时域微分性质，可得

$$\sum_{i=0}^{n} a_i (j\omega)^i R(\omega) = \sum_{j=0}^{m} b_j (j\omega)^j E(\omega) \tag{3-77}$$

可见，通过傅里叶变换，可以把常系数线性微分方程变成关于激励和响应的傅里叶变换的代数方程，从而使问题得以简化，于是得出系统的频率响应

$$H(\omega) = \frac{R(\omega)}{E(\omega)} = \frac{\sum_{j=0}^{m} b_j (j\omega)^j}{\sum_{i=0}^{n} a_i (j\omega)^i} \tag{3-78}$$

上式表明，$H(\omega)$只与系统本身有关，而与激励无关。

Example 3-15: Consider a stable LTI system that is characterized by the differential equation

$$\frac{d^2 r(t)}{dt^2} + 4\frac{dr(t)}{dt} + 3r(t) = \frac{de(t)}{dt} + 2e(t)$$

determine the frequency response and the corresponding impulse response.

Solution: Taking the Fourier transform on both sides, we get

$$(j\omega)^2 R(\omega) + 4j\omega R(\omega) + 3R(\omega) = j\omega E(\omega) + 2E(\omega)$$

therefore,

$$H(\omega) = \frac{R(\omega)}{E(\omega)} = \frac{(j\omega) + 2}{(j\omega)^2 + 4(j\omega) + 3}$$

To determine the corresponding impulse response, we use the method of partial fraction to obtain the following expressions for $H(\omega)$

$$H(\omega) = \frac{j\omega + 2}{(j\omega + 1)(j\omega + 3)} = \frac{A_1}{j\omega + 1} + \frac{A_2}{j\omega + 3}$$

and

$$A_1 = \left.\frac{j\omega + 2}{j\omega + 3}\right|_{j\omega = -1} = \frac{1}{2}, \quad A_2 = \left.\frac{j\omega + 2}{j\omega + 1}\right|_{j\omega = -3} = \frac{1}{2}$$

The inverse transform of each term can be recognized by inspection with the result that

$$h(t) = \left(\frac{1}{2}e^{-t} + \frac{1}{2}e^{-3t}\right)u(t)$$

(2) 单位冲激响应描述的系统（the unit impulse response representation of system）

这种情况下已知系统的单位冲激响应$h(t)$，然后对单位冲激响应求傅里叶变换即可。

Example 3-16: Suppose that the unit impulse response of a system $h(t) = 2e^{-2t}u(t)$, find the frequency response of the system.

Solution: According to the definition of the Fourier transform, we have

$$H(j\omega) = \int_{-\infty}^{\infty} 2e^{-2t}u(t)e^{-j\omega t}dt$$

$$= \int_{0}^{\infty} 2e^{-2t}e^{-j\omega t}dt$$

$$= \left.\frac{-2}{2+j\omega}e^{-(2+j\omega)t}\right|_{0}^{\infty} = \frac{2}{2+j\omega}$$

If we have known the system function, then we can also determine the differential equation of the system.

For example:

$$H(\omega) = \frac{j\omega + 2}{(j\omega)^2 + 4(j\omega) + 3}$$

because
$$H(\omega) = \frac{R(\omega)}{E(\omega)} = \frac{j\omega + 2}{(j\omega)^2 + 4(j\omega) + 3}$$
then
$$(j\omega)^2 R(\omega) + 4j\omega R(\omega) + 3R(\omega) = j\omega E(\omega) + 2E(\omega)$$
therefore, the differential equation of the system is
$$\frac{d^2 r(t)}{dt^2} + 4\frac{dr(t)}{dt} + 3r(t) = \frac{de(t)}{dt} + 2e(t)$$

3.6.2 系统的频域分析(Frequency Domain Analysis of System)

时域法求系统响应时，要遇到如何求卷积积分这样一个数学问题。根据傅里叶变换的时域卷积定理，若对式(3-73)两端求傅里叶变换，则有
$$R(\omega) = E(\omega)H(\omega)$$
对其求傅里叶反变换得
$$r(t) = \mathscr{F}^{-1}[R(\omega)] = \mathscr{F}^{-1}[E(\omega)H(\omega)] \tag{3-79}$$

应用上式求解系统零状态响应的方法实质上就是系统的频域分析方法。频域分析方法将时域中的卷积运算变换成频域的相乘关系，这给系统响应的求解带来很大方便。

The method of Eq. (3-79) on solving the zero state response of the system is essentially a frequency domain analysis method. Frequency domain analysis method transforms the convolution in time domain into the product in frequency domain, which brings great convenience to solve the system response.

由信号 $e(t)$ 的傅里叶变换可知，任意信号 $e(t)$ 可以表示为无穷多个虚指数信号 $e^{j\omega t}$ 的线性组合，即
$$e(t) = \frac{1}{2\pi}\int_{-\infty}^{\infty} E(\omega)e^{j\omega t}d\omega$$

式中，各个虚指数信号 $e^{j\omega t}$ 的系数大小可以看作是 $\frac{E(\omega)d\omega}{2\pi}$，即任意信号 $e(t)$ 是由无穷多个基本信号 $e^{j\omega t}$ 组合而成，那么为了求信号 $e(t)$ 激励下系统的零状态响应，首先分析在基本信号 $e^{j\omega t}$ 激励下系统的零状态响应 $r_1(t)$。

设线性时不变系统的单位冲激响应为 $h(t)$，则系统对基本信号 $e^{j\omega t}$ 的零状态响应为
$$r_1(t) = e^{j\omega t} * h(t)$$
根据卷积的定义有
$$\begin{aligned} r_1(t) &= e^{j\omega t} * h(t) = \int_{-\infty}^{\infty} h(\tau)e^{j\omega(t-\tau)}d\tau \\ &= e^{j\omega t}\int_{-\infty}^{\infty} h(\tau)e^{-j\omega\tau}d\tau \\ &= e^{j\omega t}H(\omega) \end{aligned} \tag{3-80}$$

上式表明，一个线性时不变系统，对基本信号 $e^{j\omega t}$ 的零状态响应是基本信号 $e^{j\omega t}$ 本身乘以一个与时间无关的常系数 $H(\omega)$，而 $H(\omega)$ 正是该系统单位冲激响应 $h(t)$ 的傅里叶变换。

Eq. (3-80) shows that for a linear time-invariant system, the zero state response to basic signal $e^{j\omega t}$ is that the signal itself is multiplied by a constant factor $H(\omega)$, and $H(\omega)$ just is the Fourier transform of the impulse response.

由于任意信号 $e(t)$ 可以表示为无穷多个基本信号 $e^{j\omega t}$ 的线性组合，因而应用线性叠加性质不难得出任意信号 $e(t)$ 激励下系统的零状态响应。其推导过程如下：

$$e^{j\omega t} \to H(\omega)e^{j\omega t}$$

根据线性特性可得

$$\frac{1}{2\pi}E(\omega)e^{j\omega t}d\omega \to \frac{1}{2\pi}E(\omega)H(\omega)e^{j\omega t}d\omega$$

$$\int_{-\infty}^{\infty}\frac{1}{2\pi}E(\omega)e^{j\omega t}d\omega \to \int_{-\infty}^{\infty}\frac{1}{2\pi}E(\omega)H(\omega)e^{j\omega t}d\omega$$

所以

$$e(t) \to r(t) = \mathscr{F}^{-1}[E(\omega)H(\omega)]$$

由此可得用频域分析方法求解系统零状态响应的步骤为：

第一步，求激励信号 $e(t)$ 的傅里叶变换 $E(\omega)$；

第二步，求系统的频率响应函数 $H(\omega)$；

第三步，求零状态响应 $r(t)$ 的傅里叶变换 $R(\omega) = E(\omega)H(\omega)$；

第四步，求 $R(\omega)$ 的傅里叶逆变换，即可得到 $r(t) = \mathscr{F}^{-1}[E(\omega)H(\omega)]$。

Example 3-17: Consider the system of Example 3-15, and suppose that the input is $e(t) = e^{-t}u(t)$, determine the zero state response of the system.

Solution:
$$R(\omega) = H(\omega) \cdot E(\omega) = \frac{j\omega + 2}{(j\omega + 1)(j\omega + 3)} \cdot \frac{1}{j\omega + 1}$$

In this case the partial fraction expansion takes the form

$$R(\omega) = \frac{A_{11}}{j\omega + 1} + \frac{A_{12}}{(j\omega + 1)^2} + \frac{A_2}{j\omega + 3}$$

where A_{11}, A_{12}, A_2 are constants to be determined,

$$A_{11} = \frac{d[(j\omega + 1)^2 R(\omega)]}{dj\omega}\bigg|_{j\omega = -1} = \frac{j\omega + 3 - j\omega - 2}{(j\omega + 3)^2}\bigg|_{j\omega = -1} = \frac{1}{4}$$

$$A_{12} = (j\omega + 1)^2 R(\omega)\big|_{j\omega = -1} = \frac{j\omega + 2}{j\omega + 3}\bigg|_{j\omega = -1} = \frac{1}{2}$$

$$A_2 = (j\omega + 3)R(\omega)\big|_{j\omega = -3} = \frac{j\omega + 2}{(j\omega + 1)^2}\bigg|_{j\omega = -3} = -\frac{1}{4}$$

so that
$$R(\omega) = \frac{1/4}{j\omega + 1} + \frac{1/2}{(j\omega + 1)^2} - \frac{1/4}{j\omega + 3}$$

the inverse transform is then found to be

$$r(t) = \left(\frac{1}{4}e^{-t} + \frac{1}{2}te^{-t} - \frac{1}{4}e^{-3t}\right)u(t)$$

由例 3-17 可知，利用频域分析法能够对系统的零状态响应求解。该方法的优点是时域的卷积运算变为频域的代数运算，代价是求解正反两次傅里叶变换。

3.7 无失真传输系统（Distortionless Transmission System）

信号经系统传输，要受到系统函数 $H(\omega)$ 的加权，输出波形发生了变化，与输入波形不同，即产生失真(distortion)。在信号传输过程中，为了不丢失信息，系统应该不失真地传输信号。信号失真有以下两类：

一类是线性失真(linear distortion)，包括两方面。振幅失真是系统对信号中各频率分量的幅度产生不同程度的衰减(放大)，使各频率分量之间的相对振幅关系发生了变化；相位失真是系统对信号中各频率分量产生的相移与频率不成正比，使各频率分量在时间轴上的相对位置发生了变化。这两种失真都不会使信号产生新的频率分量(without new frequency components produced)。

另一类是非线性失真(nonlinear distortion)，是由信号通过非线性系统产生的，特点是信号通过系统后产生了新的频率分量(without new frequency components produced)。

在工程设计中往往要求系统要无失真传输，所谓无失真传输是指响应信号与激励信号相比，只有幅度的大小和出现时间的先后不同，而没有波形上的变化。

Distortionless transmission is that compared with the exciting signal, the response of the system has only the difference of amplitude and time, but no change on the waveform.

设激励信号为 $e(t)$，系统的响应为 $r(t)$，则系统无失真时，系统输出为

$$r(t) = Ke(t - t_d) \tag{3-81}$$

其中，$K \neq 0$ 为系统的增益，t_d 为延迟时间，两者都为常数。满足此条件时，$r(t)$ 波形是 $e(t)$ 经 t_d 时间的滞后，虽然幅度有系数 K 倍的变化，但波形形状不变。

由式(3-81)可以得到无失真传输系统的时域无失真条件，一是幅度倍乘 K，二是波形滞后 t_d。对式(3-81)两边同时进行傅里叶变换可得

$$R(\omega) = KE(\omega)e^{-j\omega t_d}$$

因此

$$H(\omega) = \frac{R(\omega)}{E(\omega)} = Ke^{-j\omega t_d} \tag{3-82}$$

其幅频特性和相频特性分别为

$$\begin{cases} |H(\omega)| = K \\ \varphi(\omega) = -\omega t_d \end{cases} \tag{3-83}$$

式(3-82)和式(3-83)就是无失真系统的频域无失真条件。欲使信号通过线性时不变系统时不产生任何失真，必须在信号的全部频带内，要求系统频率响应的幅频特性为与频率无关的常数 K，相频特性与 ω 成正比，是一条过原点的负斜率直线。

Eq. (3-82) and Eq. (3-83) imply that for no distortion at the output of a linear time-invariant system, two requirements must be satisfied: ① The amplitude response is flat. ② The phase response is a linear function of frequency. The two requirements can be shown in Figure 3-23.

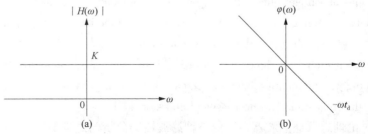

图 3-23 无失真传输系统的幅频和相频特性

从物理概念上也可以对无失真传输条件得到直观地解释。由于频率响应的幅度 $|H(\omega)|$ 为常数 K，响应中各频率分量幅度的相对大小将与激励信号的情况一样，因而没有幅度失真。

要保证没有相位失真，必须使响应中各频率分量与激励中各对应分量滞后同样的时间，这一要求反映到相频特性就是一条过原点的直线。

When the first condition is satisfied, it is said that there is no amplitude distortion. When the second condition is satisfied, there is no phase distortion. For distortionless transmission both conditions must be satisfied.

The second requirement is often specified in an equivalent way using the time delay. Define the time delay of the system by

$$T_d(\omega) = \frac{d\varphi(\omega)}{d\omega} \quad (3-84)$$

for distortionless transmission, the time delay is

$$T_d(\omega) = \frac{d\varphi(\omega)}{d\omega} = -t_d = \text{constant}$$

If $T_d(\omega)$ is not a constant, then there will be phase distortion because the phase response $\varphi(\omega)$ is not a linear function of frequency.

例 3-18： 已知某连续时间 LTI 系统的频率响应为

$$H(\omega) = \frac{1-j\omega}{1+j\omega}$$

试求该系统的幅度响应 $|H(\omega)|$ 和相位响应 $\varphi(\omega)$，并判断该系统是否为无失真传输系统。

解： $|H(\omega)| = \left|\dfrac{1-j\omega}{1+j\omega}\right| = 1$，$\varphi(\omega) = -\arctan\omega - \arctan\omega = -2\arctan\omega$

由于该系统的幅度响应 $|H(\omega)|$ 对所有的频率都为常数，这类系统被称为全通系统。但由于该系统的相位响应 $\varphi(\omega)$ 不是 ω 的线性函数，因此该系统不是无失真传输系统。

3.8 理想低通滤波器（Ideal Lowpass Filter）

所谓滤波器是指能够有选择性地让输入信号中某些频率分量通过，而其他频率分量通过很少的连续系统。在实际应用中，按照允许通过的频率成分划分，滤波器可分为低通、高通、带通和带阻等几种，它们在理想情况下的幅度响应分别如图 3-24 所示，其中 ω_c 是低通、高通的截止频率，ω_1 和 ω_2 是带通和带阻的截止频率。

Filters are a class of filters specifically intended to accurately or approximately select some bands of frequencies and reject others. In the practical application, several basic types of filter are widely used and have been given names indicative of their function. For example, lowpass filter, highpass filter, bandpass filter and band rejection filter and so on. The amplitude responses of continuous-time ideal filters are shown in Figure 3-24. In each case, the cutoff frequencies are the frequencies defining the boundaries between frequencies that are passed and frequencies that are rejected—i. e., the frequencies in the passband and stopband.

本节重点讨论理想低通滤波器，其他三种滤波器的分析与之类似。

理想低通滤波器的幅度响应 $|H(\omega)|$ 在通带 $0 \sim \omega_c$ 内恒为 1，在通带之外为 0；相位响应 $\varphi(\omega)$ 在通带内与 ω 成线性关系，如图 3-25 所示，其频率响应可表示为

$$H(\omega) = \begin{cases} e^{-j\omega t_d}, & \omega \leq |\omega_c| \\ 0, & \omega > |\omega_c| \end{cases} \tag{3-85}$$

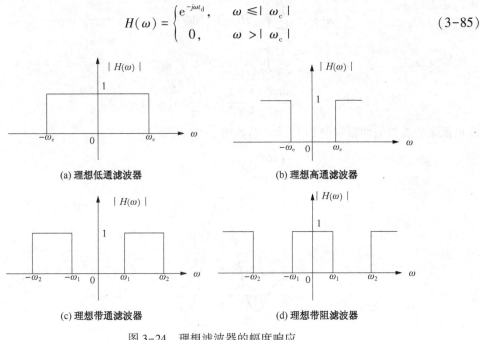

图 3-24 理想滤波器的幅度响应

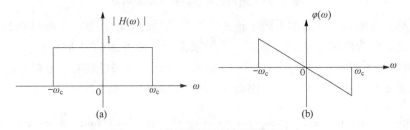

图 3-25 理想低通滤波器的频率响应

As can be seen from Eq. (3-85) or from Figure 3-25, ideal lowpass filters have perfect frequency selectivity. That is, they pass without attenuation all frequencies at or lower than the cutoff frequency ω_c and completely stop all frequencies in the stopband (i.e., higher than ω_c). Moreover, these filters have zero phase characteristics, so they introduce no phase distortion.

由于理想低通滤波器的通频带不是无穷大而是有限值,故也称为带限系统(band limited system)。显然,信号通过这种带限系统时,将会产生失真,失真的大小一方面取决于带限系统的通带宽度(band width),另一方面也取决于输入信号的有效带宽,这就是信号与系统的频率匹配(frequency matching)概念。由此可见,理想低通滤波器的通带宽窄是相对于输入信号的有效带宽而言的,当理想低通滤波器的通带宽度大于所要传输的信号有效带宽时,就可以认为系统的频带足够宽,信号通过时也能实现无失真传输。

下面分析冲激信号和阶跃信号通过理想低通滤波器的响应,这些响应的特点具有普遍意义,可以得到一些有用的结论。

3.8.1 理想低通滤波器的冲激响应(Impulse Response of the Ideal Lowpass Filter)

系统的冲激响应 $h(t)$ 就是当系统输入激励为冲激信号 $\delta(t)$ 时产生的输出响应,而且冲激响应 $h(t)$ 与系统函数 $H(\omega)$ 是一对傅里叶变换对,因此,理想低通滤波器的冲激响应为

$$h(t) = \frac{1}{2\pi}\int_{-\infty}^{\infty} H(\omega) e^{j\omega t} d\omega$$

$$= \frac{1}{2\pi}\int_{-\omega_c}^{\omega_c} e^{j\omega(t-t_d)} d\omega$$

$$= \frac{\sin\omega_c(t-t_d)}{\pi(t-t_d)} = \frac{\omega_c}{\pi} Sa(\omega_c(t-t_d)) \tag{3-86}$$

理想低通滤波器的冲激响应 $h(t)$ 的波形如图 3-26 所示。

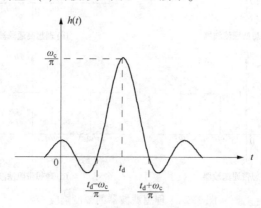

图 3-26 理想低通滤波器的冲激响应

由图可见，冲激响应的波形不同于输入的冲激信号的波形，是一个抽样函数，产生了很大的失真。这是因为理想低通滤波器是一个带限系统，而冲激信号 $\delta(t)$ 的频谱函数为常数 1，其频带宽度为无穷大。另外，由图 3-26 也可以发现，冲激响应在 $t<0$ 时也存在输出，这说明理想低通滤波器是非因果系统，因而是物理不可实现的连续系统。

We conclude from the impulse response function that the peak value of the response, $\frac{\omega_c}{\pi}$, is proportional to the cutoff frequency. Note that as $\omega_c \to \infty$ the filter becomes an all-pass filter and the peak of the impulse response $\frac{\omega_c}{\pi} \to \infty$. In other words, the impulse response approaches the input, an impulse. That is,

$$\lim_{\omega_c \to \infty} \frac{\omega_c}{\pi} Sa[\omega_c(t-t_d)] = \delta(t-t_d) \tag{3-87}$$

so that

$$\int_{-\infty}^{\infty} \frac{\omega_c}{\pi} Sa[\omega_c(t-t_d)] dt = 1$$

or

$$\int_{-\infty}^{\infty} Sa(x) dx = \pi \tag{3-88}$$

because the sampling signal $Sa(t)$ is an even function, thus,

$$\int_{-\infty}^{0} Sa(x) dx = \int_{0}^{\infty} Sa(x) dx$$

therefore,

$$\int_{-\infty}^{0} Sa(x) dx = \int_{0}^{\infty} Sa(x) dx = \frac{\pi}{2} \tag{3-89}$$

Again, the impulse response is not causal, because of $h(t) \neq 0$, $t<0$. Why is the ideal lowpass filter non-causal? Because it has finite bandwidth and linear phase characteristics.

3.8.2 理想低通滤波器的阶跃响应(Step Response of the Ideal Lowpass Filter)

如果理想低通滤波器的输入是一个单位阶跃信号 $u(t)$，则系统的输出响应称为阶跃响应，以符号 $g(t)$ 表示。由于单位阶跃信号是单位冲激信号的积分，根据线性时不变系统的特性，系统阶跃响应应是系统冲激响应的积分，即

$$g(t) = \int_{-\infty}^{t} h(\tau)\mathrm{d}\tau$$
$$= \int_{-\infty}^{t} \frac{\omega_c}{\pi} Sa[\omega_c(\tau - t_d)]\mathrm{d}\tau$$

令 $\omega_c(\tau - t_d) = x$，于是有

$$g(t) = \frac{1}{\pi}\int_{-\infty}^{\omega_c(t-t_d)} Sa(x)\mathrm{d}x$$
$$= \frac{1}{\pi}\int_{-\infty}^{0} Sa(x)\mathrm{d}x + \frac{1}{\pi}\int_{0}^{\omega_c(t-t_d)} Sa(x)\mathrm{d}x$$
$$= \frac{1}{2} + \frac{1}{\pi}\int_{0}^{\omega_c(t-t_d)} Sa(x)\mathrm{d}x \tag{3-90}$$

其波形如图 3-27 所示。

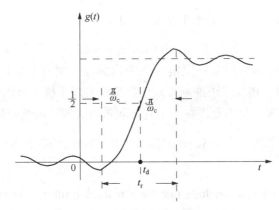

图 3-27 理想低通滤波器的阶跃响应

由图可见，阶跃响应 $g(t)$ 比输入阶跃信号 $u(t)$ 延迟一段时间。当 $t=t_d$ 时，$g(t)=0.5$，阶跃响应波形的斜率最大。若将阶跃响应从最小值上升到最大值所需要时间称为阶跃响应的上升时间 t_r，则上升时间 t_r 与理想低通滤波器的通带宽度 ω_c 成反比。

The step response $g(t)$ of the ideal lowpass filter in continuous-time is displayed in Figure 3-27. We note that the step response exhibits several characteristics that may not be desirable. In particular, for the filter, the step response overshoots its long-term final value and exhibits oscillatory behavior. The step response undergoes its most significant change in value over the interval of $\frac{2\pi}{\omega_c}$. That is, the so-called rise time of the step response, a rough measure of the response time of the filter, is inversely related to the bandwidth of the filter.

由阶跃响应的波形还可以发现，阶跃信号通过滤波器后，在其间断点的前后出现了振荡（oscillation），其振荡的最大峰值约为阶跃突变值的 9% 左右。而且如果增加滤波器的带宽（bandwidth），峰值的位置将趋于间断点，振荡起伏增多，衰减随之加快，但峰值并不减小，这种现象称为吉布斯现象（Gibbs phenomenon）。

3.8.3 佩利-维纳准则（Paley-Wiener Criterion）

由前可知，理想低通滤波器是非因果系统，在物理上是不可实现的，下面讨论物理可实现系统的约束条件。

一般来说，一个系统是否是物理可实现的，可用下面的准则来判断：

就时间域特性而言，一个物理可实现系统的单位冲激响应 $h(t)$ 应满足因果条件，即

$$t < 0 \text{ 时}, \quad h(t) = 0 \tag{3-91}$$

或表示为

$$h(t) = h(t)u(t)$$

也就是说单位冲激响应 $h(t)$ 波形的出现必须是有起因的，不能在冲激作用前就产生响应。

就频域特性而言，物理可实现系统的幅频特性应满足的条件是

$$\int_{-\infty}^{\infty} \left| \frac{\ln |H(\omega)|}{1 + \omega^2} \right| d\omega < \infty \tag{3-92}$$

且

$$\int_{-\infty}^{\infty} |H(\omega)|^2 d\omega < \infty \tag{3-93}$$

上面两个式子称为佩利-维纳准则（Paley-Wiener criterion）。

如果系统的频率响应函数在某一限定的频带内为零，则有 $|\ln|H(\omega)|| = \infty$，从而式(3-92)的积分不收敛，违反了该准则。对于物理可实现系统，可以允许 $\ln|H(\omega)|$ 在某些不连续的频率点上为零，但不允许在一个有限频带内为零。因此，理想滤波器都是物理不可实现的。对于无失真系统，由于 $\int_{-\infty}^{\infty} |H(\omega)|^2 d\omega \to \infty$，式(3-93)不成立，因而是物理不可实现的系统。

From the criterion, we can conclude that for a realizable filter, the magnitude of $H(\omega)$ may not fall off toward zero faster than a function of simple exponential order (for example, $e^{-k|\omega|}$, but not as $e^{-k\omega^2}$), and the attenuation may not be infinite over any band of frequencies of nonzero width.

For example, let $H(\omega) = e^{-\omega^2}$, then

$$\int_{-\infty}^{\infty} \frac{\ln |H(\omega)|}{1 + \omega^2} d\omega = \int_{-\infty}^{\infty} \frac{\omega^2}{1 + \omega^2} d\omega = \int_{-\infty}^{\infty} \frac{\omega^2 + 1 - 1}{\omega^2 + 1} d\omega$$

$$= \int_{-\infty}^{\infty} \left(1 - \frac{1}{\omega^2 + 1}\right) d\omega \to \infty$$

so that, $H(\omega) = e^{-\omega^2}$ cannot realize. Therefore, we often attempt to build realizable filters to approximate the ideal filter characteristic as closely as possible.

3.9 利用 MATLAB 进行连续时间信号与系统的频域分析（Fourier Analysis of Continuous-time Signals and Systems Using MATLAB）

3.9.1 傅里叶变换的 MATLAB 实现（MATLAB realization of Fourier transform）

MATLAB 的符号工具箱（Symbolic Mth Toolbox）中提供了能直接求解傅里叶变换及逆变换的函数 fourier( ) 和 ifourier( )。两者的调用格式如下。

(1) Fourier 变换
① F=fourier(f)，符号函数 f 的 Fourier 变换，默认返回关于 ω 的函数；
② F=fourier(f, v)，返回关于符号对象 v 的函数，而不是默认的 ω；
③ F=fourier(f, u, v)，对关于 u 的函数 f 进行变换，返回关于符号对象 v 的函数。

(2) Fourier 逆变换
① f=ifourier(F)，函数 F 的 Fourier 逆变换，默认返回关于 x 的函数 f；
② f=ifourier(F, u)，返回关于 u 的函数 f；
③ f=ifourier(F, v, u)，对关于 v 的函数 F 进行变换，返回关于 u 的函数 f。

注意，在调用函数 fourier( ) 和 ifourier( ) 之前，要用 syms 命令对所用到的变量进行说明，即要将这些变量说明成符号变量。

Example 3-19：Compute the Fourier transform of $f(t)=e^{-2|t|}$ using MATLAB.

Solution：Run the following MATLAB commands：

```
syms t
fourier(exp(-2*abs(t)))
```

ans =
   4/(4+w^2)

Example 3-20：Compute the inverse Fourier transform of $F(\omega)=\dfrac{1}{1+\omega^2}$ using MATLAB.

Solution：Run the following MATLAB commands：

```
syms t w
ifourier(1/(1+w^2), t)
```

ans =
   1/2*exp(t)*heaviside(-t)+1/2*exp(-t)*heaviside(t)

Example 3-21：Please plot the waveform and the amplitude spectrum of $f(t)=\dfrac{1}{2}e^{-2t}u(t)$.

Solution：Run the following MATLAB commands：

```
syms t w v x
x=1/2*exp(-2*t)*sym('Heaviside(t)')
F=fourier(x);
subplot(211);
ezplot(x)
subplot(212);
ezplot(abs(F))
```

The waveform of $f(t)$ and it's amplitude spectrum are shown in Figure 3-28.

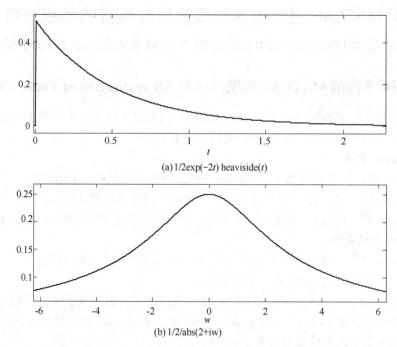

(a) 1/2exp(-2t) heaviside(t)

(b) 1/2/abs(2+iw)

Figure 3-28 the waveform of $f(t)$ and it's amplitude spectrum

Example 3-22: Giving $f(t) = Sa(t)$ and it's Fourier transform
$$F(\omega) = \pi g_2(\omega) = \pi[u(\omega + 1) - u(\omega - 1)]$$
calculate theFourier transform of the signal $f_1(t) = \pi G_2(t)$ by using MATLAB.

Solution: Run the following MATLAB commands:

```
r=0.01; t=-15:r:15;
f=sin(t)./t;
f1=pi*(Heaviside(t+1)-Heaviside(t-1));
N=500; W=5*pi*1;
k=-N:N; w=k*W/N;
F=r*sinc(t/pi)*exp(-j*t'*w);
F1=r*f1*exp(-j*t'*w);
subplot(221);
plot(t, f)
xlabel('t'); ylabel('f(t)');
subplot(222);
plot(w, F)
axis([-2 2 -1 4]);
xlabel('w'); ylabel('F(w)');
subplot(223);
plot(t, f1)
axis([-2 2 -1 4]);
xlabel('t'); ylabel('f1(t)');
```

```
subplot(224);
plot(w, F1)
axis([-20 20 -3 7]);
xlabel('w'); ylabel('F1(w)')
```

The waveform of $f(t)$ and $f_1(t)$ and their spectrum graph are shown in Figure 3-29.

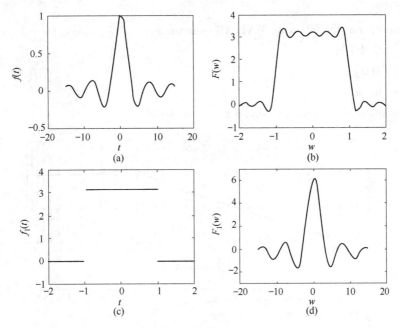

Figure 3-29 Example of symmetry properties

3.9.2　用 MATLAB 分析连续时间系统的频率特性(MATLAB Analysis for Frequency Characteristic of Continuous-time Systems)

MATLAB 提供了专门对连续系统频率响应 $H(\omega)$ 的函数 freqs( )。该函数可以求出系统频率响应的数值解,并可绘出系统的幅频及相频响应曲线。freqs( )函数有如下四种调用格式:

① **h=freqs(b, a, w)**,*b* 和 *a* 为多项式形式的 $H(\omega)$ 的分子分母系数向量,ω 对应于 $H(\omega)$ 的频率范围;

For example, run the following MATLAB commands:

```
a=[1 2 1];
b=[0 1];
w=0:0.5:2*pi;
h=freqs(b, a, w);
```

then the running results of the program are:

1.0000	0.4800-0.6400i	0-0.5000i
-0.1183-0.2840i	-0.1200-0.1600i	-0.0999-0.0951i
-0.0800-0.0600i	-0.0641-0.0399i	-0.0519-0.0277i

-0.0426-0.0199i   -0.0355-0.0148i   -0.0300-0.0113i
-0.0256-0.0088i

② [h, w]=freqs(b, a)，计算默认频率范围内 200 个频率点的系统频率响应；

③ [h, w]=freqs(b, a, n)，计算默认频率范围内 n 个频率点的系统频率响应；

④ **freqs(b, a)**，不返回系统频率响应的数值，而是以对数坐标的方式绘出系统的幅频及相频响应曲线。

For example, run the following MATLAB commands:
  a=[1 0.4 1];
  b=[1 0 0];
  freqs(b, a);
the running results of the program is shown in Figure 3-30.

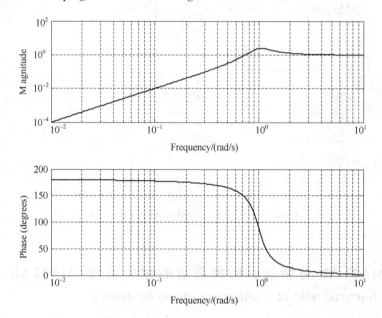

Figure 3-30 frequency response curve

3.9.3 连续信号的采样与重构(Sampling and recovery of continuous-time signal)

当采样频率 $\omega_s = 2\omega_m$ 时，称为临界采样，取 $\omega_c = \omega_m$。下列程序实现对信号 $Sa(t)$ 的采样及由该采样信号恢复 $Sa(t)$（图 3-31）。

  wm = 1; wc = wm;
  Ts = pi/wm; ws = 2 * pi/Ts;
  n = -100:100;
  nTs = n * Ts;
  f = sinc(nTs/pi);
  Dt = 0.05; t = -15:Dt:15;
  fa = f * Ts * wc/pi * sinc((wc/pi) * (ones(length(nTs), 1) * t-nTs' * ones(1, length(t))));%重构
  t1 = -15:0.5:15;

```
f1 = sinc(t1/pi);
subplot(211)
stem(t1, f1);
xlabel('kTs');
ylabel('f(kTs)');
title('Sa(t)= sinc(t/pi)的临界采样信号');
subplot(212)
plot(t, fa);
xlabel('t');
ylabel('fa(t)');
title('由 Sa(t)= sinc(t/pi)的临界采样信号重构 Sa(t)')
```

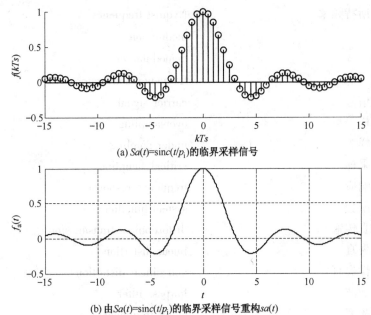

(a) $Sa(t)=sinc(t/p_1)$的临界采样信号

(b) 由$Sa(t)=sinc(t/p_1)$的临界采样信号重构$sa(t)$

图 3-31 临界采样信号及信号恢复

关键词(Key Words and Phrases)

(1) 频域分析　　　　　　　　　　frequency domain analysis
(2) 傅里叶级数　　　　　　　　　Fourier series
(3) 基波　　　　　　　　　　　　fundamental component
(4) 谐波　　　　　　　　　　　　harmonic component
(5) 频谱　　　　　　　　　　　　spectrum
(6) 幅度频谱　　　　　　　　　　amplitude spectrum
(7) 相位频谱　　　　　　　　　　phase spectrum
(8) 偶函数　　　　　　　　　　　even function
(9) 奇函数　　　　　　　　　　　odd function
(10) 奇谐函数　　　　　　　　　　odd harmonic function

（11）偶谐函数	even harmonic function
（12）半波镜像函数	half-wave mirror function
（13）半波重叠函数	half-wave overlapping function
（14）频带宽度	frequency bandwidth
（15）傅里叶变换	Fourier transform
（16）频谱密度函数	spectrum density function
（17）充分条件	sufficient condition
（18）必要条件	necessary condition
（19）绝对可积	absolutely integrable
（20）抽样定理	sampling theorem
（21）奈奎斯特间隔	Nyquist interval
（22）奈奎斯特频率	Nyquist frequency
（23）调制	modulation
（24）解调	demodulation
（25）变频	frequency conversion
（26）载波信号	carrier signal
（27）过采样	oversampling
（28）欠采样	undersampling
（29）临界采样	critical sampling
（30）频率响应	frequency response
（31）系统函数	system function
（32）无失真传输	distortionless transmission
（33）线性失真	linear distortion
（34）非线性失真	non-linear distortion
（35）低通滤波器	lowpass filter
（36）截止频率	cutoff frequency
（37）通带	passband
（38）阻带	stopband
（39）带限系统	band-limited system
（40）振荡	oscillation
（41）吉布斯现象	Gibbs phenomenon
（42）佩利-维纳准则	Paley-Wiener criterion
（43）频率混叠	frequency overlap
（44）截断误差	truncation error
（45）混叠误差	overlap error

Exercises

3-1 The periodic rectangular signal is shown in Figure 3-32. Find the Fourier series (Triangular and exponential form) and plot the amplitude spectrum.

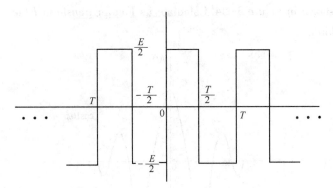

Figure 3-32 the waveform of Exercises 3-1

3-2 Expand the following signals into triangular Fourier series on the given interval.
(1) $f(t)=t$, $(-\pi, \pi)$
(2) $f(t)=e^t$, $(0, 1)$

3-3 Calculate the inverse Fourier transform for the following $F(\omega)$ is shown in Figure 3-33.

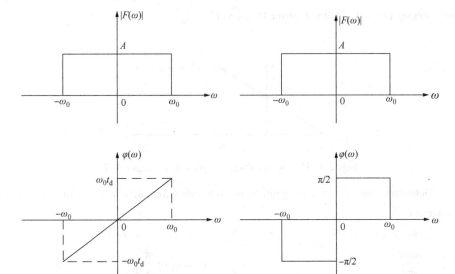

Figure 3-33 the amplitude and phase spectrum of $F(\omega)$ in Exercise 3-3

3-4 Calculate the inverse Fourier transform for each given $F(\omega)$ as follow.
(1) $F(\omega)=\delta(\omega-\omega_0)$
(2) $F(\omega)=\begin{cases}\dfrac{\omega_0}{\pi} & |\omega|\leq\omega_0 \\ 0 & \text{otherwise}\end{cases}$
(3) $F(\omega)=u(\omega+\omega_0)-u(\omega-\omega_0)$

3-5 If $f(t)\leftrightarrow F(\omega)$, find the Fourier transforms for the following signals by using properties of the Fourier transform.
(1) $t\cdot f(2t)$
(2) $(t-1)f(t)$
(3) $(t-2)f(-2t)$
(4) $f(2t-5)$
(5) $t\dfrac{df(t)}{dt}$
(6) $\int_{-\infty}^{t}\tau f(\tau)d\tau$

3-6 $f(t)$ is shown in Figure 3-34. Calculate its Fourier transform $F(\omega)$ by using properties of the Fourier transform.

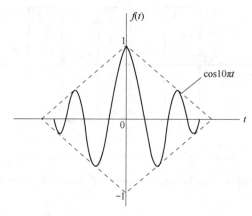

Figure 3-34　the waveform of $f(t)$ in Exercises 3-6

3-7 Determine the spectrum density function for the given signal that is shown in Figure3-35 (Using the integral property of the Fourier transform).

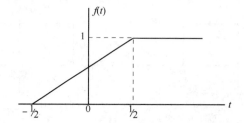

Figure 3-35　the waveform of $f(t)$ in Exercises 3-7

3-8 Determine the Fourier transforms for the following signals by using properties of the Fourier transform.

(1) $e^{-\alpha t}\cos\omega_0 t u(t)$, $\alpha>0$　　　　(2) $e^{-3t}[u(t+2)-u(t-2)]$

(3) $\int_{-\infty}^{t}\dfrac{\sin\pi\tau}{\pi\tau}d\tau$　　　　(4) $\sin t+\cos\left(t+\dfrac{\pi}{4}\right)$

3-9 Find the zero state response by using frequency domain analysis method. The system function and excitation signal is given as following.

$$H(\omega)=\dfrac{1}{j\omega+2},\ e(t)=e^{-3t}u(t)$$

3-10 Find the frequency response and the impulse response of the following systems.

(1) $e(t)=e^{-t}u(t)$, $r(t)=(e^{-2t}+e^{-3t})u(t)$
(2) $e(t)=e^{-3t}u(t)$, $r(t)=e^{-3(t-2)}u(t-2)$

3-11 Determine the frequency response and the impulse response for the systems described by the following differential equations.

(1) $\dfrac{dr(t)}{dt}+3r(t)=e(t)$

(2) $\dfrac{d^2r(t)}{dt^2}+5\dfrac{dr(t)}{dt}+6r(t)=\dfrac{-de(t)}{dt}$

3-12 In order to distortionless transmission for the given system (or circuit), as shown in Figure 3-12, determine the relationship between R_1, R_2 and C_1, C_2.

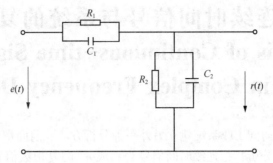

Figure 3-36   the circuit diagram of Exercises 3-12

3-13 The system function of ideal lowpass filler is

$$H(\omega) = \begin{cases} 1 & |\omega| \leq \dfrac{2\pi}{\tau} \\ 0 & |\omega| < \dfrac{2\pi}{\tau} \end{cases}$$

and the Fourier transform of the input signal $E(\omega) = \tau S_a \dfrac{\omega \tau}{2}$. Determine the zero state response $r(t)$ by using the convolution theorem.

3-14 For each of the following differential equations and excitations for causal and stable LTI systems, determine the zero state responses of these systems using the frequency domain analysis method.

(1) $\dfrac{d^2 r(t)}{dt^2} + 5\dfrac{dr(t)}{dt} + 6r(t) = \dfrac{de(t)}{dt}$, $e(t) = e^{-t} u(t)$

(2) $\dfrac{d^2 r(t)}{dt^2} + 3\dfrac{dr(t)}{dt} + 2r(t) = \dfrac{1}{2}\dfrac{de(t)}{dt} + 2e(t)$, $e(t) = u(t)$

(3) $\dfrac{d^2 r(t)}{dt^2} + 4\dfrac{dr(t)}{dt} + 3r(t) = \dfrac{de(t)}{dt} + 2e(t)$, $e(t) = e^{-2t} u(t)$

3-15 The output $r(t)$ of a causal LTI system is related to the input $e(t)$ by the differential equation

$$\dfrac{dr(t)}{dt} + 2r(t) = e(t)$$

(1) Determine the frequency response of the system.

(2) If $e(t) = e^{-t} u(t)$, determine $R(\omega)$ and the output $r(t)$.

(3) Repeat part (2), if the input has as its Fourier transform $E(\omega) = \dfrac{1+j\omega}{2+j\omega}$.

3-16 Given the system function of a system $H(\omega) = \dfrac{j\omega}{-\omega^2 + 5j\omega + 6}$, the initial conditions $r(0^-) = 2$, $r'(0^-) = 1$, the excitation signal $e(t) = e^{-t} u(t)$. Find the complete response of the system.

第4章 连续时间信号与系统的复频域分析
(Analysis of Continuous-time Signals and Systems in Complex Frequency Domain)

本章我们将对连续时间 LTI 系统建立另外一种分析方法。前面章节中我们研究了连续时间信号与系统的傅里叶分析,把连续时间信号表示成 $e^{j\omega t}$ 的复指数信号的线性组合就构成了傅里叶级数和傅里叶变换的基础,傅里叶分析在涉及信号和 LTI 系统的众多领域使用非常广泛。当我们把 $e^{j\omega t}$ 扩展成复变量 s,且 $s=\sigma+j\omega$ 的复指数函数,也就是将连续时间信号看成是一系列 e^{st} 复指数叠加之和,这样就由傅里叶变换推广至拉普拉斯变换(the Laplace transform),简称拉氏变换。拉普拉斯变换可以理解为傅里叶变换的推广形式,而傅里叶变换则是拉普拉斯变换当 s 取 $s=j\omega$ 时的特殊情况。

利用拉普拉斯变换方法可以将 LTI 系统的时域模型简便地进行变换,在 s 域中求解系统的响应,再还原成响应的时间函数。拉氏变换分析方法作为研究连续信号与 LTI 系统的有力工具,在电学和力学等众多领域都有广泛的应用。

In this chapter we will introduce the tool of the Laplace transform for analysis of the continuous-time signals and systems. In the preceding chapter, we have seen that the Fourier transform of a signal $f(t)$ is a representation of the function as a continuous sum of exponential function of the form $e^{j\omega t}$. The Laplace transform can be viewed as an extension of Fourier transform in which we represent a function $f(t)$ as linear combinations of complex exponentials of the form of e^{st} with $s=\sigma+j\omega$, where $s=\sigma+j\omega$ is a complex frequency.

4.1 拉普拉斯变换(The Laplace Transform)

本节将从傅里叶变换的定义推导出拉普拉斯变换的定义,并讨论拉氏变换的收敛条件,介绍拉氏变换的收敛域(region of convergence)。

4.1.1 拉普拉斯变换的定义(Definition of the Laplace Transform)

前一章我们学到,傅里叶变换存在的充分条件是狄利克雷条件(Dirichlet conditions),但我们经常遇到很多信号都不满足狄利克雷条件,很难从傅里叶变换定义直接求出,如单位阶跃信号;还有一些信号,如实指数信号 $f(t)=e^{at}(a>0)$,其傅里叶变换均不存在。为了使更多的信号存在变换,简化运算分析过程,我们引入一个衰减因子 $e^{-\sigma t}$,将不满足狄利克雷条件的信号 $f(t)$ 与衰减因子 $e^{-\sigma t}$ 相乘,使 $f(t)e^{-\sigma t}$ 的积分得以收敛,即 $\int_{-\infty}^{\infty} |f(t)e^{-\sigma t}| \mathrm{d}t < \infty$ 满足绝对可积的条件,则其傅里叶变换存在。

$$\mathscr{F}[f(t)e^{-\sigma t}] = \int_{-\infty}^{\infty} f(t)e^{-\sigma t}e^{-j\omega t}\mathrm{d}t = \int_{-\infty}^{\infty} f(t)e^{-(\sigma+j\omega)t}\mathrm{d}t \tag{4-1}$$

我们将上式的积分结果看作是 $(\sigma+j\omega)$ 的函数,记为 $F(\sigma+j\omega)$,则上式变为

$$F(\sigma + j\omega) = \int_{-\infty}^{\infty} [f(t)e^{-\sigma t}] e^{-j\omega t} dt = \int_{-\infty}^{\infty} f(t) e^{-(\sigma + j\omega)t} dt$$

令 $s = \sigma + j\omega$，上式可写为

$$F(s) = \int_{-\infty}^{\infty} f(t) e^{-st} dt \qquad (4-2)$$

式(4-2)中 $F(s)$ 称为信号 $f(t)$ 的拉普拉斯变换。相应的信号 $f(t)e^{-\sigma t}$ 的傅里叶反变换为

$$e^{-\sigma t} f(t) = \frac{1}{2\pi} \int_{-\infty}^{\infty} F(\sigma + j\omega) e^{j\omega t} d\omega \qquad (4-3)$$

上式两边同乘以 $e^{\sigma t}$ 可以得到

$$f(t) = \frac{1}{2\pi} \int_{-\infty}^{\infty} F(\sigma + j\omega) e^{(\sigma + j\omega)t} d\omega \qquad (4-4)$$

同样，令 $s = \sigma + j\omega$，则 $d\omega = \frac{1}{j} ds$，代入上式，将积分变量改为 s，则式(4-4)可以写为

$$f(t) = \frac{1}{2\pi j} \int_{\sigma - j\infty}^{\sigma + j\infty} F(s) e^{st} ds \qquad (4-5)$$

式(4-5)中 $f(t)$ 称为 $F(s)$ 的拉普拉斯反变换。此式也表明信号 $f(t)$ 可以分解为复指数 e^{st} 的线性组合。

我们将式(4-2)和式(4-5)称为双边拉普拉斯变换(the bilateral Laplace transform or the two-sided Laplace transform)的一对变换式。式(4-2)为双边拉氏变换正变换式，式(4-5)为双边拉氏变换反变换式，可表示为

$$F(s) = \mathscr{L}[f(t)]$$
$$f(t) = \mathscr{L}^{-1}[F(s)]$$
$$f(t) \overset{\mathscr{L}}{\leftrightarrow} F(s)$$

Notice that we have multiplied a given function $f(t)$ by $e^{-\sigma t}$, where σ is any real number. Thus the convergence of the resulting Fourier integral is greatly enhanced by the so-called convergence factor $e^{-\sigma t}$. From this viewpoint, the Fourier transform is a special case in which $s = j\omega$.

我们可以看到当 $s = j\omega$ 时，式(4-2)变为 $F(j\omega) = \int_{-\infty}^{\infty} f(t) e^{-j\omega t} dt$，也就是信号 $f(t)$ 的傅里叶变换，可写作 $F(s)|_{s=j\omega} = \mathscr{F}[f(t)]$。

4.1.2　拉普拉斯变换的收敛域 (Region of Convergence for the Laplace Transform)

从式(4-2)可以看出双边拉氏变换是在傅里叶变换的被积函数中增加了衰减因子 $e^{-\sigma t}$，适当地选取 σ 的值，可使被积函数收敛，因而双边拉氏变换存在。由傅里叶变换存在的条件，我们可以得出拉氏变换存在的充分条件是 $f(t)e^{-\sigma t}$ 绝对可积，即

$$\int_{-\infty}^{\infty} |f(t)e^{-\sigma t}| dt < \infty \qquad (4-6)$$

我们将使得拉氏变换存在的 σ 的取值范围称为收敛域(Region of Convergence, ROC)。

The range of values of σ (or Re s) for which the integral Eq. (4-6) converges is referred to as the region of convergence (we abbreviate as *ROC*) of the Laplace transform.

Example 4-1: Find the Laplace Transform of the signal $f(t) = e^{-at} u(t)$ (it's Fourier transform

$F(\omega)$ converges for $a>0$).

Solution: Because $F(\omega) = \int_{-\infty}^{\infty} e^{-at} u(t) e^{-j\omega t} = \int_{0}^{\infty} e^{-at} e^{-j\omega t} dt = \dfrac{1}{a+j\omega}$, $a > 0$ its Laplace transform can be found by equation (4-2),

$$F(s) = \int_{-\infty}^{\infty} e^{-at} u(t) e^{-st} dt = \int_{0}^{\infty} e^{-(s+a)t} dt$$

Let $s = \sigma + j\omega$,

$$F(\sigma + j\omega) = \int_{0}^{\infty} e^{-(\sigma+a)t} \cdot e^{-j\omega t} dt$$

and we get

$$F(\sigma + j\omega) = \dfrac{1}{(\sigma + a) + j\omega}, \quad \sigma + a > 0$$

therefore,

$$F(s) = \dfrac{1}{s + a}, \quad \sigma > -a$$

that is,

$$e^{-at} u(t) \overset{\mathscr{L}}{\leftrightarrow} \dfrac{1}{s + a}, \quad \sigma = (\text{Re } s) > -a \tag{4-7}$$

In Eq. (4-7), the Laplace transform converges only for $\sigma > -a$. If a is positive, then $F(s)$ can be evaluated at $\sigma = 0$ to obtain $F(0+j\omega) = \dfrac{1}{a+j\omega}$. We can see that the Laplace transform is equal to the Fourier transform for $\sigma = 0$. If a is negative or zero, the Laplace transform still exists, but the Fourier transform does not.

Example 4-2: Let us consider another example of the function $f(t) = -e^{-at} u(-t)$, please compare with Example 4-1.

The Laplace transform of the signal can be found by Eq. (4-2):

$$F(s) = -\int_{-\infty}^{\infty} e^{-at} e^{-st} u(-t) dt$$

$$= -\int_{-\infty}^{0} e^{-(s+a)t} dt = \int_{0}^{-\infty} e^{-(s+a)t} dt$$

$$= \dfrac{1}{s+a} \int_{-\infty}^{0} e^{-(s+a)t} d[-(s+a)]t$$

$$= \dfrac{1}{s+a} e^{-(s+a)t} \Big|_{-\infty}^{0} = \dfrac{1}{s+a} e^{-(\sigma+a)t} e^{-j\omega t} \Big|_{-\infty}^{0}$$

$$= \dfrac{1}{s+a}(1 - 0) = \dfrac{1}{s+a}, \text{ if and only if } \sigma = (\text{Re } s) < -a$$

that is,

$$-e^{-at} u(-t) \overset{\mathscr{L}}{\leftrightarrow} \dfrac{1}{s+a}, \quad \sigma = (\text{Re } s) < -a \tag{4-8}$$

Comparing the two examples, the Laplace transform of both the signals have the same algebraic expression. But we also see that the set of values of σ for which the expression is valid is different in the two examples. Therefore, we must specify both the algebraic expression and the range of values of σ for which this expression is valid for the Laplace transform of a signal.

可见，求信号的双边拉氏变换时，要同时给出收敛域。任意信号和它的双边拉氏变换连同收敛域才是一一对应的。

例 4-3：求有始有终信号 $f_1(t)=\delta(t)$ 和 $f_2(t)=u(t)-u(t-2)$ 的双边拉氏变换。

解：① $F_1(s)=\int_{-\infty}^{\infty}f(t)\mathrm{e}^{-st}\mathrm{d}t=\int_{-\infty}^{\infty}\delta(t)\mathrm{e}^{-st}\mathrm{d}t=\int_{-\infty}^{\infty}\delta(t)\mathrm{d}t=1$

$-\infty<\sigma=(\mathrm{Re}\,s)<\infty$，收敛域为整个 s 平面。

② $F_2(s)=\int_{-\infty}^{\infty}f(t)\mathrm{e}^{-st}\mathrm{d}t=\int_0^2\mathrm{e}^{-st}\mathrm{d}t=\dfrac{1}{s}\int_2^0\mathrm{e}^{-st}\mathrm{d}(-st)\dfrac{1}{s}\mathrm{e}^{-st}\bigg|_2^0=\dfrac{1}{s}(1-\mathrm{e}^{-2s})$

$-\infty<\sigma=(\mathrm{Re}\,s)<\infty$，收敛域也为整个 s 平面。

例 4-4：求无始无终函数 $f(t)=\mathrm{e}^{\alpha t}u(t)-\mathrm{e}^{\beta t}u(-t)$ 的拉氏变换。

解：$F(s)=\int_{-\infty}^{\infty}f(t)\mathrm{e}^{-st}\mathrm{d}t=\int_0^{\infty}\mathrm{e}^{\alpha t}\mathrm{e}^{-st}\mathrm{d}t-\int_{-\infty}^0\mathrm{e}^{\beta t}\mathrm{e}^{-st}\mathrm{d}t$

$=\dfrac{1}{s-\alpha}+\dfrac{1}{s-\beta}$

由公式(4-7)，我们可以得出因果信号 $\mathrm{e}^{\alpha t}u(t)$ 的收敛域满足 $\sigma=(\mathrm{Re}\,s)>\alpha$。

由公式(4-8)，我们可以得出反因果信号 $-\mathrm{e}^{\beta t}u(-t)$ 的收敛域满足 $\sigma=(\mathrm{Re}\,s)<\beta$。

当 $\alpha<\beta$ 时，双边信号 $f(t)$ 的拉氏变换存在，其收敛域为 $\alpha<\sigma=(\mathrm{Re}\,s)<\beta$。

当 $\alpha>\beta$ 时，没有公共的收敛域，其拉氏变换不存在。

由上面的几个例子可得到下面几点结论：

① 有始无终函数(如因果信号 $f(t)u(t)$)的双边拉氏变换如果存在，则收敛域为 $\sigma>\sigma_c$，见图 4-1；

② 无始有终函数(如反因果信号 $f(t)u(-t)$)的双边拉氏变换如果存在，则收敛域为 $\sigma<\sigma_c$，见图 4-2；

③ 无始无终函数的双边拉氏变换如果存在，则收敛域为 $\sigma_{c1}>\sigma>\sigma_{c2}$，见图 4-3；

④ 有始有终函数的双边拉氏变换如果存在，则收敛域为整个 s 平面；

⑤ 收敛域的边界总是平行于 $j\omega$ 轴的经过 $F(s)$ 极点的竖直线。

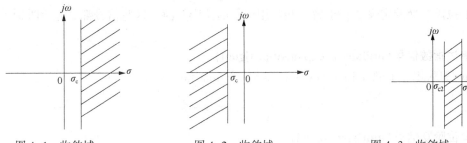

图 4-1 收敛域 $\sigma>\sigma_c$ 图 4-2 收敛域 $\sigma<\sigma_c$ 图 4-3 收敛域 $\sigma_{c2}<\sigma<\sigma_{c1}$

A signal either does not have a Laplace transform or falls into one of the four categries covered by the above examples. Thus, for any signal with a Laplace transform, the ROC must be the entire s-plane (for finite-length signals), a left-half plane (for left-sided signals), a right-half plane (for right-sided signals), or a signal strip (for two-sided signals).

4.1.3 单边拉普拉斯变换(The Unilateral Laplace Transform)

我们使用的信号 $f(t)$ 都有起始时刻，一般通常取起始时刻为 $t=0$，那么 $t<0$ 时，$f(t)=$

0，即信号 $f(t)$ 为因果信号。我们取因果信号 $f(t)$ 的拉氏变换，则式(4-2)和式(4-5)可写为

$$F(s) = \mathscr{L}[f(t)] = \int_0^\infty f(t)\mathrm{e}^{-st}\mathrm{d}t \tag{4-9}$$

$$f(t) = \mathscr{L}^{-1}[F(s)] = \frac{1}{2\pi j}\int_{\sigma-j\infty}^{\sigma+j\infty} F(s)\mathrm{e}^{st}\mathrm{d}s \quad t>0 \tag{4-10}$$

式(4-9)和式(4-10)为单边拉普拉斯变换(the unilateral Laplace transform or the one-sided Laplace transform)的一对变换式。单边拉氏变换对于分析具有初始条件的常系数线性微分方程描述的连续因果系统具有重要意义。

在这里我们注意两点，一是信号的选取，信号 $f(t)$ 一般为因果信号；二是积分下限的选取，在本书中选取积分下限为 0^- 时刻，这主要考虑到 $f(t)$ 中有可能包含冲激信号等奇异函数，且从 s 域分析系统时我们很容易求解出系统的零输入响应。为了简单起见，我们常将积分下限写为 0，在这里作以说明。

在工程实际中所遇到的信号大多是因果信号，对因果信号 $f(t)$，其单边拉氏变换与双边拉氏变换相等。考虑到工程实用性，本章主要讨论单边拉氏变换。

The unilateralLaplace transform is of considerable value in analyzing causal systems which specified by linear constant-coefficient differential equation with nonzero initial condition. Hence in this book the development is limited to the unilateral(one-sided) Laplace transform, which will be referred to as the Laplace transform.

从上面对收敛域的讨论可知，对有始有终的因果信号 $f(t)$，其单边拉氏变换的收敛域为整个 s 平面；对有始无终的因果信号 $f(t)$，其单边拉氏变换的收敛域为 $\sigma>\sigma_{\max}$，$\sigma_{\max}$ 是 $F(s)$ 极点的最大实部，也就是说单边拉氏变换的收敛域可根据 $F(s)$ 的极点写出。为简单起见，在后面的学习中，有时在给出 $F(s)$ 的同时，不再给出收敛域，特此说明。

4.2 常用信号的拉普拉斯变换（Some Laplace Transform Pairs）

本节由拉氏变换的定义式，求解一些常用信号的拉氏变换，这些信号都是起始时刻为零的因果信号。

(1) 单边指数信号(unilateral exponential signal)

We have already calculated the key transform pair from Eq. (4-7)

$$\mathrm{e}^{-at}u(t) \overset{\mathscr{L}}{\leftrightarrow} \frac{1}{s+a}, \quad \sigma>-a$$

(2) 单位阶跃信号(unit step signal)

For the unilateral exponential signal, when $a=0$, we can get the transform pair of the unit step signal：

$$u(t) \overset{\mathscr{L}}{\leftrightarrow} \frac{1}{s}, \quad \sigma>0 \tag{4-11}$$

(3) 正弦型信号(sinusoidal signal)

If $a=j\omega$ or $a=-j\omega$ in the unilateral exponential signal $\mathrm{e}^{-at}u(t)$, then from Eq. (4-7),

$$\mathrm{e}^{-j\omega t}u(t) \overset{\mathscr{L}}{\leftrightarrow} \frac{1}{s+j\omega}, \quad \sigma>0 \tag{4-12}$$

$$e^{+j\omega t}u(t) \xleftrightarrow{\mathscr{L}} \frac{1}{s-j\omega}, \quad \sigma > 0 \qquad (4-13)$$

Adding Eq. (4-12) and Eq. (4-13) yield the transform pair for the cosine function,

$$\cos\omega t u(t) \xleftrightarrow{\mathscr{L}} \frac{1}{2} \cdot \frac{1}{s+j\omega} + \frac{1}{2}\frac{1}{s-j\omega} = \frac{s}{s^2+\omega^2}, \quad \sigma > 0 \qquad (4-14)$$

Subtracting Eq. (4-13) from Eq. (4-12) yield the transform pair for the sinusoidal function,

$$\sin\omega t u(t) \xleftrightarrow{\mathscr{L}} \frac{1}{2j}\left(\frac{1}{s+j\omega} - \frac{1}{s-j\omega}\right) = \frac{\omega}{s^2+\omega^2}, \quad \sigma > 0 \qquad (4-15)$$

Suppose that, $a = \sigma \pm j\omega$ in Eq. (4-7), then the resulting transform pair is

$$e^{-(\sigma \pm j\omega)t}u(t) \xleftrightarrow{\mathscr{L}} \frac{1}{s+(\sigma \pm j\omega)}$$

if we break $a = \sigma \pm j\omega$ into real and imaginary parts, we obtain two transform pairs

$$e^{-\sigma t}e^{\mp j\omega t}u(t) = e^{-\sigma t}[\cos\omega t \mp j\sin\omega t] \xleftrightarrow{\mathscr{L}} \frac{(s+\sigma)\pm j\omega}{(s+\sigma)^2 \pm \omega^2} = \frac{s+\sigma}{(s+\sigma)^2+\omega^2} \pm j\frac{\omega}{(s+\sigma)^2+\omega^2}$$

or

$$e^{-at}\cos\omega t u(t) \xleftrightarrow{\mathscr{L}} \frac{s+a}{(s+a)^2+\omega^2}, \quad \sigma > -a \qquad (4-16)$$

$$e^{-at}\sin\omega t u(t) \xleftrightarrow{\mathscr{L}} \frac{\omega}{(s+a)^2+\omega^2}, \quad \sigma > -a \qquad (4-17)$$

(4) 单位冲激信号(unit impulse signal)

According to the definition of the bilateral Laplace transform, we have

$$F(s) = \int_{-\infty}^{\infty} \delta(t)e^{-st}dt = 1$$

the ROC is the entire s-plane, and denoted by

$$\delta(t) \xleftrightarrow{\mathscr{L}} 1, \quad -\infty < \sigma < \infty \qquad (4-18)$$

(5) t 的正幂次信号 $t^n u(t)$ (n 为正整数) (positive power signal of t)

Finally, using the bilateral Laplace transform to find $L[t^n u(t)]$.

$$\mathscr{L}[t^n u(t)] = \int_{-\infty}^{\infty} t^n e^{-st}u(t)dt = -\frac{1}{s}\int_0^{\infty} t^n de^{-st}$$

$$= -\frac{1}{s}t^n e^{-st}\Big|_0^{\infty} + \frac{n}{s}\int_0^{\infty} t^{n-1}e^{-st}dt$$

because the first term is zero when $\sigma > 0$, thus,

$$\mathscr{L}[t^n u(t)] = 0 + \frac{n}{s}\int_0^{\infty} t^{n-1}e^{-st}dt = \frac{n}{s}\mathscr{L}[t^{n-1}u(t)]$$

according to the above reasoning, we can conclude

$$\mathscr{L}[t^n u(t)] = \frac{n}{s} \cdot \frac{n-1}{s}\mathscr{L}[t^{n-2}]$$

$$= \frac{n}{s} \cdot \frac{n-1}{s} \cdot \cdots \cdot \frac{1}{s}\mathscr{L}[t^0]$$

$$= \frac{n!}{s^{n+1}}, \quad \sigma > 0$$

that is,
$$t^n u(t) \overset{\mathscr{L}}{\leftrightarrow} \frac{n!}{s^{n+1}}, \ \sigma > 0 \quad (4-19)$$

when $n = 1$,
$$tu(t) \overset{\mathscr{L}}{\leftrightarrow} \frac{1}{s^2}, \ \sigma > 0 \quad (4-20)$$

when $n = 2$,
$$t^2 u(t) \overset{\mathscr{L}}{\leftrightarrow} \frac{2}{s^3}, \ \sigma > 0 \quad (4-21)$$

下面将常用信号的傅里叶变换列于表 4-1，以便查阅。

表 4-1 常用信号的拉氏变换

序号	时域表示 $f(t)$	拉氏变换 $F(s)$	收敛域
1	$e^{-at}u(t)$	$\frac{1}{s+a}$	$\sigma > -a$
2	$-e^{-at}u(-t)$	$\frac{1}{s+a}$	$\sigma < -a$
3	$u(t)$	$\frac{1}{s}$	$\sigma > 0$
4	$-u(-t)$	$\frac{1}{s}$	$\sigma < 0$
5	$\sin\omega t u(t)$	$\frac{\omega}{s^2+\omega^2}$	$\sigma > 0$
6	$\cos\omega t u(t)$	$\frac{s}{s^2+\omega^2}$	>0
7	$e^{-at}\sin\omega t u(t)$	$\frac{\omega}{(s+a)^2+\omega^2}$	$\sigma > -a$
8	$e^{-at}\cos\omega t u(t)$	$\frac{s+a}{(s+a)^2+\omega^2}$	$\sigma > -a$
9	$\delta(t)$	1	$-\infty < \sigma < \infty$
10	$t^n u(t)$	$\frac{n!}{s^{n+1}}$	$\sigma > 0$
11	$e^{-at}tu(t)$	$\frac{1}{(s+a)^2}$	$\sigma > -a$
12	$e^{-at}t^2 u(t)$	$\frac{2}{(s+a)^3}$	$\sigma > -a$

以上我们用拉氏变换的定义式求出几个常用信号的拉氏变换，对于较复杂的信号，直接使用拉氏变换定义式求积分较困难，这时使用常用信号的拉氏变换和拉氏变换的性质来求取更为简便有效，下一节将讨论拉氏变换的性质。

4.3 拉普拉斯变换的性质（Properties of the Laplace Transform）

拉氏变换是傅里叶变换的推广形式，因此，两种变换的性质存在很多相似之处，由傅里叶变换可知当信号在时域有所变化时，在变换域必然有相应的体现，下面讨论的拉氏变换性质就体现了这些变化规律。另外使用拉氏变换的性质来求取信号的拉氏变换对于较复杂信号更为简便。下面主要介绍单边拉氏变换的一些基本性质。

The bilateral Laplace transform and the unilateral Laplace transform have a number of important properties, most of which are the same. A particularly important difference between the properties of the bilateral and unilateral Laplace transforms is the differentiation property. The properties of the Laplace transform are so similar to the Fourier transform. The derivations of many of these results are analogous to those of the corresponding properties for the Fourier transform. So we will not present the derivations in detail.

4.3.1 线性(Linearity)

若信号 $f_1(t)$ 与 $f_2(t)$ 的拉氏变换分别为 $F_1(s)$ 与 $F_2(s)$，其收敛域 σ 分别位于 R_1 与 R_2 的取值范围。a_1 和 a_2 为任意常数。简记为：

$$f_1(t) \overset{\mathscr{L}}{\leftrightarrow} F_1(s) \quad ROC:R_1$$

$$f_2(t) \overset{\mathscr{L}}{\leftrightarrow} F_2(s) \quad ROC:R_2$$

则有

$$a_1 f_1(t) + a_2 f_2(t) \overset{\mathscr{L}}{\leftrightarrow} a_1 F_1(s) + a_2 F_2(s) \quad ROC:R_c \supseteq R_1 \cap R_2 \tag{4-22}$$

这一性质可由拉氏变换的定义直接得到证明。线性叠加信号的拉氏变换其收敛域一般是原信号拉氏变换收敛域的重叠部分。值得注意的是，如果原来信号拉氏变换的收敛域没有重叠部分，那么由线性叠加后的新信号将不存在拉氏变换；若两个信号经过线性运算得到的信号是一个时限信号，则其收敛域变为整个 s 平面。

As indicated, the region of convergence of $F(s)$ is at least the intersection of R_1 and R_2, which could be empty, in which case $F(s)$ has no region of convergence—i.e., $f(t)$ has no Laplace transform. The ROC can also be larger than the intersection.

例 4-5：求信号 $f(t)=u(t)-u(t-2)$ 的双边拉氏变换。

解：方法一：由信号的拉氏变换定义式，我们可以得到

$$F(s) = \int_{-\infty}^{\infty} f(t) e^{-st} dt$$

$$= \int_0^2 e^{-st} dt = -\frac{1}{s} e^{-st} \Big|_0^2 = \frac{1}{s}(1 - e^{-2s}), \quad -\infty < \sigma < \infty$$

方法二：由常用信号的拉氏变换对，可知

$$u(t) \overset{\mathscr{L}}{\leftrightarrow} \frac{1}{s}, \quad \sigma > 0$$

$$u(t-2) \overset{\mathscr{L}}{\leftrightarrow} \frac{1}{s} e^{-2s}, \quad \sigma > 0$$

$$u(t) - u(t-2) \overset{\mathscr{L}}{\leftrightarrow} \frac{1}{s} - \frac{1}{s} e^{-2s}, \quad -\infty < \sigma < \infty$$

因为由单位阶跃信号和其时移信号线性组合的新信号 $f(t)=u(t)-u(t-2)$ 为时限信号，因此收敛域为整个 s 平面。

Example 4-6：Find the Laplace transform of the DC signal $f(t)=1$.

Solution: The DC signal $f(t)=1$ can be written as $f(t)=u(t)+u(-t)$, from Eq. (4-7) and Eq. (4-8),

$$u(t) \overset{\mathscr{L}}{\leftrightarrow} \frac{1}{s}, \quad \sigma > 0$$

$$u(-t) \overset{\mathscr{L}}{\leftrightarrow} \frac{-1}{s}, \quad \sigma < 0$$

There is no common region of convergence [the region of convergence of $F(s)$ is empty], and thus $f(t)$ has no Laplace transform.

4.3.2 时移特性(Time Shifting)

若 $f(t)u(t) \overset{\mathscr{L}}{\leftrightarrow} F(s)$, $ROC: R$, 则

$$f(t-t_0)u(t-t_0) \overset{\mathscr{L}}{\leftrightarrow} F(s)e^{-st_0}, \quad ROC: R_c = R \tag{4-23}$$

证明：由拉氏变换的定义有

$$\mathscr{L}[f(t-t_0)u(t-t_0)] = \int_0^\infty f(t-t_0)u(t-t_0)e^{-st}dt$$

$$= \int_{t_0}^\infty f(t-t_0)e^{-st}dt$$

变量代换，令 $t-t_0=x$, 则 $dt=dx$, $0<x<\infty$

上式变为
$$\mathscr{L}[f(t-t_0)u(t-t_0)] = \int_0^\infty f(x)e^{-s(x-t_0)}dx$$

$$= e^{-st_0}\int_0^\infty f(x)e^{-st}dx = e^{-st_0}F(s)$$

由上面推导可知，只要 $F(s)$ 存在，则 $e^{-st_0}F(s)$ 也存在，且收敛域与 $F(s)$ 相同。

同理，不难推导出信号双边拉氏变换的时移性质，证明方法相同。

若 $f(t) \overset{\mathscr{L}}{\leftrightarrow} F(s)$, $ROC: R$, 则有

$$f(t-t_0) \overset{\mathscr{L}}{\leftrightarrow} F(s)e^{-st_0}, \quad ROC: R_c = R \tag{4-24}$$

Example 4-7: Find the Laplace transform of $f(t)=\frac{E}{T}t[u(t)-u(t-T)]$.

Solution: $f(t)$ can be represented as

$$f(t) = \frac{E}{T}t[u(t) - u(t-T)] = \frac{E}{T}tu(t) - \frac{E}{T}tu(t-T)$$

$$= \frac{E}{T}[tu(t) - (t-T+T)u(t-T)]$$

$$= \frac{E}{T}[tu(t) - (t-T) - Tu(t-T)]$$

From the Eq. (4-20) and Eq. (4-23), then

$$\mathscr{L}[tu(t)] = \frac{1}{s^2}$$

$$\mathscr{L}[(t-T)u(t-T)] = \frac{1}{s^2}e^{-sT}$$

and

$$\mathscr{L}[Tu(t-T)] = \frac{T}{s}e^{-sT}$$

hence,

$$\mathscr{L}[f(t)] = \frac{E}{T}\left[\frac{1}{s^2} - \frac{1}{s^2}e^{-sT} - \frac{T}{s}e^{-sT}\right]$$

$$= \frac{E}{s^2T}[1 - e^{-sT} - Tse^{-sT}]$$

4.3.3 复频移特性(Shifting in the s-domain)

若 $f(t) \overset{\mathscr{L}}{\leftrightarrow} F(s)$, $ROC:R$, 则有：

$$e^{s_0 t}f(t) \overset{\mathscr{L}}{\leftrightarrow} F(s-s_0), \quad ROC:R_c = R + \sigma_0 \tag{4-25}$$

式中，s_0 为任意复常数，上式表明时域信号 $f(t)$ 乘以 $e^{s_0 t}$，相当于其变换式在 s 域平移 s_0。式(4-25)所表示的 s 域平移特性，对双边拉氏变换也同样适用。

Example 4-8: Determine the Laplace transform of the signal $f(t) = e^{-at}\sin\omega_0 t u(t)$.

Solution: Because we have known

$$\sin\omega_0 t u(t) \overset{\mathscr{L}}{\leftrightarrow} \frac{\omega_0}{s^2 + \omega_0^2}, \quad \sigma > 0$$

Using Eq. (4-25), we get

$$\mathscr{L}[e^{-at}\sin\omega_0 t u(t)] = \frac{\omega_0}{(s+a)^2 + \omega_0^2}, \quad \sigma > -a$$

4.3.4 尺度变换特性(Time Scaling)

若 $f(t) \overset{\mathscr{L}}{\leftrightarrow} F(s)$, $ROC:R$, 则有

$$f(at) \overset{\mathscr{L}}{\leftrightarrow} \frac{1}{a}F\left(\frac{s}{a}\right), \quad \text{且}(a > 0), \quad ROC:aR \tag{4-26}$$

式中，$a>0$ 是为了保证 $f(at)$ 仍然为因果信号。若 $F(s)$ 的 ROC 为 $(\mathrm{Re}\, s)>\sigma_0$，则 $F\left(\dfrac{s}{a}\right)$ 的 ROC 为 $\mathrm{Re}\left\{\dfrac{s}{a}\right\}>\sigma_0$，即 $(\mathrm{Re}\, s)>a\sigma_0$。对于双边拉氏变换，$a$ 的取值可正可负，其 ROC 的变化与单边拉氏变换相同，即：

若 $f(t) \overset{\mathscr{L}}{\leftrightarrow} F(s)$, $ROC:R$, 则有

$$f(at) \overset{\mathscr{L}}{\leftrightarrow} \frac{1}{|a|}F\left(\frac{s}{a}\right) \quad ROC:aR \tag{4-27}$$

Example 4-9: We have known the Laplace transform $\mathscr{L}[e^{-t}u(t)] = \dfrac{1}{s+1}$, $(\sigma>-1)$, please find the Laplace transform $\mathscr{L}[f(at)]$.

Solution: $\mathscr{L}[f(at)] = \dfrac{1}{a} \cdot \dfrac{1}{\dfrac{s}{a}+1} = \dfrac{1}{s+a} \quad (\sigma > -a)$

Example 4-10: If $f(t) \overset{\mathscr{L}}{\leftrightarrow} F(s)$, $ROC:\sigma>\sigma_0$, please find the Laplace transform $\mathscr{L}[e^{-3t}$

$f(2t+1)$].

Solution: Because $f(t) \overset{\mathscr{L}}{\leftrightarrow} F(s)$, $ROC: \sigma > \sigma_0$, using Eq. (4-24), Eq. (4-25), Eq. (4-26), we have

$$f(t+1) \overset{\mathscr{L}}{\leftrightarrow} F(s)e^s, \quad ROC: \sigma > \sigma_0$$

$$f(2t+1) \overset{\mathscr{L}}{\leftrightarrow} \frac{1}{2}F\left(\frac{s}{2}\right)e^{\frac{s}{2}}, \quad ROC: \sigma > 2\sigma_0$$

$$e^{-3t}f(2t+1) \overset{\mathscr{L}}{\leftrightarrow} \frac{1}{2}F\left(\frac{s+3}{2}\right)e^{\frac{s+3}{2}}, \quad ROC: \sigma > 2\sigma_0 - 3$$

4.3.5 时域微分特性(Differentiation in the Time Domain)

The properties of the bilateral and unilateral Laplace transforms are almost the same. But the particularly important difference between the properties of the bilateral and unilateral Laplace transforms is the differentiation property.

(1) 双边拉氏变换的时域微分特性(differentiation in the time domain of the bilateral Laplace transform)

若 $f(t) \overset{\mathscr{L}}{\leftrightarrow} F(s)$, $ROC: R$, 则有

$$\frac{df(t)}{dt} \overset{\mathscr{L}}{\leftrightarrow} sF(s), \quad ROC: R_c \supseteq R \tag{4-28}$$

(2) 单边拉氏变换的时域微分特性(differentiation in the time domain of the unilateral Laplace transform)

The differentiation property of the unilateral Laplace transform is of considerable value in analyzing causal systems which specified by linear constant-coefficient differential equation with nonzero initial condition.

若 $f(t) \overset{\mathscr{L}}{\leftrightarrow} F(s)$, $ROC: R$, 则有

$$f'(t) \overset{\mathscr{L}}{\leftrightarrow} sF(s) - f(0^-) \quad ROC: R_c \supseteq R \tag{4-29}$$

推广到一般式有

$$\frac{d^n f(t)}{dt^n} \overset{\mathscr{L}}{\leftrightarrow} s^n F(s) - s^{n-1}f(0^-) - s^{n-2}f'(0^-) - \cdots - f^{(n-1)}(0^-) \tag{4-30}$$

证明：由单边拉氏变换定义

$$\mathscr{L}\left[\frac{df(t)}{dt}\right] = \int_{0^-}^{\infty} \frac{df(t)}{dt} e^{-st} dt = \int_{0^-}^{\infty} df(t) e^{-st}$$

对上式应用分部积分法，可得

$$\mathscr{L}[f'(t)] = f(t)e^{-st}\big|_{0^-}^{\infty} + s\int_{0^-}^{\infty} f(t)e^{-st}dt$$

$$= f(t)e^{-st}\big|_{t=\infty} - f(0^-) + sF(s) = sF(s) - f(0^-)$$

设 $y(t) = \dfrac{df(t)}{dt}$，则有 $Y(s) = sF(s) - f(0^-)$。

再次应用微分性质，

$$\frac{dy(t)}{dt} = \frac{d^2f(t)}{dt^2} \overset{\mathscr{L}}{\leftrightarrow} sY(s) - y(0^-) = s^2F(s) - f'(0^-) - sf(0^-) \qquad (4\text{-}31)$$

再次应用微分性质，

$$\frac{d^3f(t)}{dt^3} \overset{\mathscr{L}}{\leftrightarrow} = s^2F(s) - s^2f(0^-) - sf'(0^-) - f'(0^-) \qquad (4\text{-}32)$$

以此类推，重复利用微分性质，可得到高阶导数的拉斯变换：

$$\frac{d^nf(t)}{dt^n} \overset{\mathscr{L}}{\leftrightarrow} s^nf(s) - s^{n-1}f(0^-) - s^{n-2}f'(0^-) - \cdots - sf^{(n-2)}(0^-) - f^{(n-1)}(0^-) \qquad (4\text{-}33)$$

If $f(t)$ is discontinuous at $t=0$ or if $f(t)$ possess an impulse or derivative of an impulse located at $t=0$, it is necessary to take the unilateral Laplace transform to be at $t=0^-$.

4.3.6 复频域微分特性(Differentiation in the s-Domain)

若 $f(t) \overset{\mathscr{L}}{\leftrightarrow} F(s)$, $ROC:R$, 则有

$$-tf(t) \overset{\mathscr{L}}{\leftrightarrow} \frac{dF(s)}{ds} \quad ROC:R_c = R \qquad (4\text{-}34)$$

$$(-t)^nf(t) \overset{\mathscr{L}}{\leftrightarrow} \frac{d^nF(s)}{ds^n} \quad ROC:R_c = R \qquad (4\text{-}35)$$

证明：由单边拉氏变换定义 $F(s) = \mathscr{L}[f(t)] = \int_0^\infty f(t)e^{-st}dt$,

上式对 s 求导，并交换微分与积分顺序，可得

$$\frac{dF(s)}{ds} = \int_0^\infty f(t)\frac{d}{ds}(e^{-st})dt = \int_0^\infty -tf(t)e^{-st}dt$$

也就是 $\mathscr{L}[-tf(t)] = \frac{dF(s)}{ds}$，记为

$$-tf(t) \overset{\mathscr{L}}{\leftrightarrow} \frac{dF(s)}{ds}$$

重复使用式(4-34)，可得

$$\frac{d^nF(s)}{ds^n} = \int_0^\infty (-t)^nf(t)e^{-st}dt$$

即

$$(-t)^nf(t) \overset{\mathscr{L}}{\leftrightarrow} \frac{d^nF(s)}{ds^n}$$

由上面的推导，我们可以看出双边拉氏变换 s 域微分特性与单边拉氏变换相同。

Example 4-11: Find the Laplace transform of $f(t) = te^{-\lambda t}u(t)$.

Solution: Method 1　Because $\mathscr{L}[tu(t)] = \frac{1}{s^2}(\sigma>0)$

using the property of shifting in the s-domain, we get

$$\mathscr{L}[te^{-\lambda t}u(t)] = \frac{1}{(s+\lambda)^2} \quad (\sigma > \lambda)$$

Method 2: By the property of differentiation in the s-domain

because
$$\mathscr{L}[e^{-\lambda t}u(t)] = \frac{1}{s+\lambda} \quad (\sigma > \lambda)$$

thus
$$\mathscr{L}[te^{-\lambda t}u(t)] = -\frac{d}{ds}\left[\frac{1}{s+\lambda}\right] = \frac{1}{(s+\lambda)^2} \quad (\sigma > \lambda) \tag{4-36}$$

4.3.7 时域卷积特性(Time Convolution Property)

若 $f_1(t) \overset{\mathscr{L}}{\leftrightarrow} F_1(s)$, $ROC:R_1$; $f_2(t) \overset{\mathscr{L}}{\leftrightarrow} F_2(s)$, $ROC:R_2$, 则有

$$\mathscr{L}[f_1(t) * f_2(t)] = F_1(s)F_2(s) \quad ROC:R_c \supseteq R_1 \cap R_2 \tag{4-37}$$

证明:
$$\mathscr{L}[f_1(t) * f_2(t)] = \int_0^\infty \left[\int_{-\infty}^\infty f_1(\tau)f_2(t-\tau)d\tau\right]e^{-st}dt$$
$$= \int_0^\infty f_1(\tau)\left[\int_0^\infty f_2(t-\tau)e^{-st}dt\right]d\tau$$
$$= \int_0^\infty f_1(\tau)F_2(s)e^{-s\tau}d\tau$$
$$= F_1(s)F_2(s)$$

上式表明两个信号时域卷积的拉氏变换等于两个信号各自拉氏变换的乘积，其收敛域包含了 $F_1(s)$ 与 $F_2(s)$ 收敛域的交集。利用时域卷积特性可以简便的由 S 域求解系统的零状态响应。

As we saw in Chapter 3, the convolution property in the context of the Laplace transform plays an important role in the analysis of linear time-invariant systems.

例 4-12：试求图 4-4 所示三角脉冲信号的拉普拉斯变换。

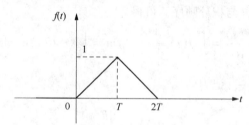

图 4-4 三角脉冲信号

解：根据卷积积分的特性，可知该三角脉冲信号可等价为两个矩形脉冲信号的卷积，即

$$f(t) = f_1(t) * f_2(t) = \frac{1}{\sqrt{T}}[u(t) - u(t-T)] * \frac{1}{\sqrt{T}}[u(t) - u(t-T)]$$

其中, $f_1(t) = f_2(t) = \frac{1}{\sqrt{T}}[u(t) - u(t-T)]$, 对应的拉普拉斯变换为

$$F_1(s) = F_2(s) = \frac{1}{\sqrt{T}}\frac{1 - e^{-Ts}}{s}$$

根据时域卷积定理，信号 $f(t) = f_1(t) * f_2(t)$ 的拉普拉斯变换为

$$F(s) = F_1(s)F_2(s) = \frac{1}{Ts^2}(1 - e^{-Ts})^2$$

4.3.8 时域积分特性(Integration in the Time Domain)

若 $f(t) \overset{\mathscr{L}}{\leftrightarrow} F(s)$, $ROC:R$, 则有

$$\int_{0^-}^{t} f(\tau)\mathrm{d}\tau \overset{\mathscr{L}}{\leftrightarrow} \frac{F(s)}{s} \quad ROC:R_c \supseteq R \cap \{\sigma > 0\} \qquad (4-38)$$

$$\int_{-\infty}^{t} f(\tau)\mathrm{d}\tau \overset{\mathscr{L}}{\leftrightarrow} \frac{f^{(-1)}(0^-)}{s} + \frac{F(s)}{s} \quad ROC:R_c \supseteq R \cap \{\sigma > 0\} \qquad (4-39)$$

式中 $f^{(-1)}(0^-) = \int_{-\infty}^{0^-} f(\tau)\mathrm{d}\tau$，是 $f(t)$ 在 $t=0^-$ 时的取值。t 取 0^- 时刻是考虑到积分式 $t=0$ 时刻可能有跳变。

证明：$\int_{-\infty}^{t} f(\tau)\mathrm{d}\tau = \int_{-\infty}^{0^-} f(\tau)\mathrm{d}\tau + \int_{0^-}^{t} f(\tau)\mathrm{d}\tau = f^{(-1)}(0^-) + \int_{0^-}^{t} f(\tau)\mathrm{d}\tau$

上式第一项 $f^{(-1)}(0^-)$ 为函数 $f(t)$ 积分的初始值(为一常数)，所以

$$\mathscr{L}[f^{(-1)}(0^-)] = \frac{f^{(-1)}(0^-)}{s}$$

上式第二项的拉氏变换，利用分部积分法，可得

$$\int_0^\infty \left[\int_0^t f(\tau)\mathrm{d}\tau\right] \mathrm{e}^{-st} \mathrm{d}t = \left[-\frac{\mathrm{e}^{-st}}{s}\int_0^t f(\tau)\mathrm{d}\tau\right]_0^\infty + \frac{1}{s}\int_0^\infty f(t)\mathrm{e}^{-st}\mathrm{d}t$$

$$= \frac{1}{s}\int_0^t f(t)\mathrm{e}^{-st}\mathrm{d}t = \frac{F(s)}{s}$$

例 4-13：试利用拉氏变换的时域积分特性求图 4-4 所示三角脉冲信号的拉普拉斯变换。

解：当信号波形可由分段函数描述时，信号导数的波形通常都较为简单，适合应用积分性质。图 4-4 所示的三角脉冲信号可以用分段函数描述为

$$f(t) = \frac{1}{T}t[u(t) - u(t-T)] + \left(2 - \frac{1}{T}t\right)[u(t-T) - u(t-2T)]$$

对上式求导，可得

$$\frac{\mathrm{d}f(t)}{\mathrm{d}t} = \frac{1}{T}[u(t) - u(t-T)] - \frac{1}{T}[u(t-T) - u(t-2T)]$$

由时移单位阶跃函数的拉普拉斯变换可知，上式的拉普拉斯变换对为

$$\frac{\mathrm{d}f(t)}{\mathrm{d}t} \leftrightarrow F_1(s) = \frac{1}{T}\frac{1-\mathrm{e}^{-Ts}}{s} - \frac{1}{T}\frac{\mathrm{e}^{-Ts}-\mathrm{e}^{-2Ts}}{s} = \frac{1}{Ts}(1-\mathrm{e}^{-Ts})^2$$

注意到三角脉冲信号是一个因果信号，所以根据积分性质可知 $f(t)$ 的拉普拉斯变换为

$$F(s) = \frac{F_1(s)}{s} = \frac{1}{Ts^2}(1-\mathrm{e}^{-Ts})^2$$

4.3.9 初值定理(Initial-value Theorem)

若 $f(t)$ 为因果信号，即 $t<0$ 时 $f(t)=0$。且 $f(t)$ 在 $t=0$ 时不包含冲激函数及其各阶导数，$f(t)$ 的拉氏变换为 $F(s)$，则有

$$f(0^+) = \lim_{s \to \infty} sF(s) \qquad (4-40)$$

式中 $F(s)$ 为有理真分式。

证明：由单边拉氏变换的时域微分性质，可得

$$sF(s) - f(0^-) = \mathscr{L}\left[\frac{\mathrm{d}f(t)}{\mathrm{d}t}\right]$$

$$= \int_{0^-}^{\infty} \frac{df(t)}{dt} e^{-st} dt$$

$$= \int_{0^-}^{0^+} \frac{df(t)}{dt} e^{-st} dt + \int_{0^+}^{\infty} \frac{df(t)}{dt} e^{-st} dt$$

$$= f(0^+) - f(0^-) + \int_{0^+}^{\infty} \frac{df(t)}{dt} e^{-st} dt$$

当 $s \to \infty$，有 $\lim\limits_{s \to \infty} \int_{0^+}^{\infty} \frac{df(t)}{dt} e^{-st} = 0$，因此

$$f(0^+) = \lim_{s \to \infty} sF(s)$$

Example 4-14: Supposing $F(s) = \dfrac{s}{(s+a)^2 + \omega^2}$, find the initial value of $f(t)$.

Solution: $f(0^+) = \lim\limits_{s \to \infty} sF(s) = \lim\limits_{s \to \infty} \dfrac{s^2}{(s+a)^2 + \omega^2} = 1$

Also, from the Laplace transform $F(s)$, we know, the time signal $f(t) = e^{-at} \cos\omega t$. So that $f(0^+) = 1$.

Example 4-15: Suppose that $F(s) = \dfrac{s}{s+1} (\sigma > -1)$, find the initial value of $f(t)$.

Solution: Because $F(s)$ is not the proper fraction, the initial-value theorem can not directly used. Hence

$$F(s) = \frac{s}{s+1} = 1 - \frac{1}{s+1} = 1 + F_1(s)$$

$$F_1(s) = -\frac{1}{s+1}$$

then

$$f(0^+) = \lim_{s \to \infty} sF_1(s) = \lim_{s \to \infty} \frac{-1}{s+1} = -1$$

however, the time domain $f(t)$ is

$$f(t) = \mathscr{L}^{-1}\left[\frac{s}{s+1}\right] = \delta(t) - e^{-t} u(t)$$

$$f(0^+) = \lim_{t \to 0^+}[\delta(t) - e^{-t} u(t)] = \delta(0^+) - e^0 u(0^+) = -1$$

The initial-value property permits one to calculate $f(0^+)$ directly from the transform $F(s)$ without the need of inverting the transform.

4.3.10 终值定理(Final-value Theorem)

若 $f(t)$ 为因果信号，且 $\lim\limits_{t \to \infty} f(t)$ 存在，$f(t)$ 的拉氏变换为 $F(s)$，则有

$$\lim_{t \to \infty} f(t) = f(\infty) = \lim_{s \to 0} sF(s) \tag{4-41}$$

证明：再次应用单边拉氏变换的时域微分性质，可得

$$sF(s) - f(0^-) = \int_{0^-}^{\infty} \frac{df(t)}{dt} e^{-st} dt$$

上式两边去取 $s \to 0$ 之极限，则有

$$\lim_{s\to 0}[sF(s)-f(0^-)] = \int_{0^-}^{\infty}\frac{\mathrm{d}f(t)}{\mathrm{d}t}\mathrm{d}t = f(\infty)-f(0^-) \tag{4-42}$$

若 $sF(s)$ 的收敛域包含 $j\omega$ 轴，也就是 $\lim\limits_{t\to\infty}f(t)=f(\infty)$ 存在，则式(4-41)可以写作

$$f(\infty)=\lim_{s\to 0}sF(s)$$

以上讨论的初值定理和终值定理，在比较复杂的系统分析中，用来确定初值和终值很方便，只要知道响应的拉氏变换，就可以确定初值和终值，而不必求出时域响应。但要注意初值定理和终值定理的使用条件。

The initial- and final-value theorems can be useful in checking the correctness of the Laplace transform calculations for a signal. But it should be noted that the constraint conditions of the two theorems.

最后，将上面讨论的拉氏变换的性质列于表 4-2 中，以方便查阅。

表 4-2 拉氏变换性质

序号	性质名称	时域信号	拉氏变换	收敛域
1	线性性质	$a_1f_1(t)+a_2f_2(t)$	$a_1F_1(s)+a_2F_2(s)$	$R_c \supseteq R_1 \cap R_2$
2	时移特性	$f(t-t_0)u(t-t_0)$	$F(s)\mathrm{e}^{-st_0}$	$R_c=R$
3	复频移特性	$\mathrm{e}^{s_0 t}f(t)$	$F(s-s_0)$	$R_c=R+\sigma_0$
4	尺度变换特性	$f(at)\,(a>0)$	$\dfrac{1}{a}F\left(\dfrac{s}{a}\right)$	$R_c=aR$
5	时域微分特性	$\dfrac{\mathrm{d}f(t)}{\mathrm{d}t}$	$sF(s)-f(0^-)$	$R_c \supseteq R$
		$\dfrac{\mathrm{d}^2 f(t)}{\mathrm{d}t^2}$	$s^2F(s)-f'(0^-)-sf(0^-)$	$R_c \supseteq R$
6	复频域微分	$-tf(t)$	$\dfrac{\mathrm{d}F(s)}{\mathrm{d}s}$	$R_c=R$
		$(-t)^n f(t)$	$\dfrac{\mathrm{d}^n F(s)}{\mathrm{d}s^n}$	$R_c=R$
7	时域积分特性	$\int_{0^-}^{t}f(\tau)\mathrm{d}\tau$	$\dfrac{F(s)}{s}$	$R_c \supseteq R \cap \{\sigma>0\}$
		$\int_{-\infty}^{t}f(\tau)\mathrm{d}\tau$	$\dfrac{f^{(-1)}(0^-)}{s}+\dfrac{F(s)}{s}$	$R_c \supseteq R \cap \{\sigma>0\}$
8	时域卷积特性	$f_1(t)*f_2(t)$	$F_1(s)F_2(s)$	$R_c \supseteq R_1 \cap R_2$
9	初值定理		$\lim\limits_{t\to 0^+}f(t)=f(0^+)=\lim\limits_{s\to\infty}sF(s)$	
10	终值定理		$\lim\limits_{t\to\infty}f(t)=f(\infty)=\lim\limits_{s\to 0}sF(s)$	

4.4 拉普拉斯反变换（The Inverse Laplace Transform）

无论对信号还是系统进行复频域分析，经常需要从信号的拉氏变换 $F(s)$ 变换到时间信号 $f(t)$，这就是信号的拉氏反变换。因为我们研究的信号和系统大多数都是因果的，所以本

— 123 —

节重点介绍单边拉普拉斯反变换。通常拉氏反变换有两种方法：部分分式法(partial fraction expansion method)和留数法(residue method)。

Partial fraction expansion method and residue method are the two usual methods of solving the inversion of the unilateralLaplace transform. The former applies to that $F(s)$ is the rational proper fraction, but the latter is more extensive.

It is notable that in this section we discuss the methods of solving the inversion of the **unilateral** Laplace transform.

4.4.1 部分分式法(Partial Fraction Expansion Method)

在 LTI 系统的微分方程求解问题中，拉普拉斯变换域方法会出现关于 s 的多项式之比的形式。通过部分分式展开，将 $F(s)$ 表示成已知基本函数的各个部分分式之和，再利用常用信号拉氏变换，即可获得拉普拉斯逆变换。

部分分式展开法适用于 $F(s)$ 为有理函数真分式(rational proper fraction)。通常将 $F(s)$ 表示为

$$F(s) = \frac{N(s)}{D(s)} = \frac{b_m S^m + b_{m-1} S^{m-1} + \cdots + b_1 S + b_0}{a_n S^n + a_{n-1} S^{n-1} + \cdots + a_1 S + a_0}$$

这里 $a_i(i=1, 2, \cdots, n)$ 和 $b_j(j=0, 1, \cdots, m)$ 均为实数，n、m 为整数，$F(s)$ 为分子多项式(numerator polynomial)与分母多项式(denominator polynomial)之比。在一般情况下，分母多项式的阶次 n 大于分子多项式的阶次 m，即 $n>m$，此时 $F(s)$ 为真分式。

If this is not the case, then we can always write $F(s)$ as the sum of a polynomial $Q(s)$ of degree $m-n$ plus a ratio of polynomial with the numerator degree one less than the denominator degree. The prime requirement of partial fraction expansion method is that $F(s)$ is the rational proper fraction.

In order to expand a rational fraction, we first find the roots of the denominator polynomial of the $p_1, p_2, \cdots, p_n(n>m)$. These roots are the poles.

根据 $F(s)$ 的极点(poles)情况，部分分式的展开有以下几种情况。

(1) $F(s)$ 的极点为实数单根(distinct roots)

若 $F(s)$ 为有理真分式($n>m$)，其极点为一阶实数根(无重根)，则 $F(s)$ 可分解为

$$F(s) = \frac{N(s)}{(s-p_1)(s-p_2)\cdots(s-p_n)} \quad (n>m)$$

$$= \frac{k_1}{s-p_1} + \frac{k_2}{s-p_2} + \cdots + \frac{k_n}{s-p_n} = \sum_{i=1}^{n} \frac{k_i}{s-p_i} \quad (4\text{-}43)$$

其中 $k_i(i=1, 2, \cdots n)$ 为待定系数，可在式(4-43)两边同乘以 $(s-p_i)$，再令 $s=p_i$，则有

$$k_i = (s-p_i)F(s)\big|_{s=p_i} \quad (4\text{-}44)$$

式(4-43)中 $F(s)$ 表示的是单边拉氏变换，则其收敛域均在最右侧极点的右边，所对应的信号都是因果信号。以下均将 $F(s)$ 的收敛域省略，特此说明。

由前面式(4-7)可知，$F(s)$ 的拉氏反变换为

$$\sum_{i=1}^{n} \frac{k_i}{s-p_i} \stackrel{\mathscr{L}^{-1}}{\longleftrightarrow} f(t) = \sum_{i=1}^{n} k_i e^{p_i t} u(t) \quad (4\text{-}45)$$

Example 4-16: Find the inverse transform of the function

$$F(s) = \frac{5(s+1)}{s(s+2)(s+4)}$$

Solution: $F(s)$ can be expanded as

$$F(s) = \frac{k_1}{s} + \frac{k_2}{s+2} + \frac{k_3}{s+4}$$

where,

$$k_1 = \frac{5(s+1)}{(s+2)(s+4)}\bigg|_{s=0} = \frac{5}{8}$$

$$k_2 = \frac{5(s+1)}{s(s+4)}\bigg|_{s=-2} = \frac{5}{4}$$

$$k_3 = \frac{5(s+1)}{s(s+2)}\bigg|_{s=-4} = \frac{-15}{8}$$

thus,

$$F(s) = \frac{\frac{5}{8}}{s} + \frac{\frac{5}{4}}{s+2} + \frac{-\frac{15}{8}}{s+4}$$

then,

$$f(t) = \mathscr{L}^{-1}[F(s)] = \mathscr{L}^{-1}\left[\frac{\frac{5}{8}}{s}\right] + \mathscr{L}^{-1}\left[\frac{\frac{5}{4}}{s+2}\right] + \mathscr{L}^{-1}\left[\frac{-\frac{15}{8}}{s+4}\right]$$

$$= \frac{5}{8}u(t) + \frac{5}{4}e^{-2t}u(t) + \left(-\frac{15}{8}\right)e^{-4t}u(t)$$

$$= \frac{5}{8}(1 + 2e^{-2t} - 3e^{-4t})u(t)$$

Example 4-17: Find the inverse transform of the function

$$F(s) = \frac{2s^2 + 6s + 6}{s^2 + 3s + 2}$$

Solution: Because $F(s)$ is not the proper fraction, $F(s)$ can be expanded as

$$F(s) = \frac{2s^2 + 6s + 6}{s^2 + 3s + 2} = 2 + \frac{2}{s^2 + 3s + 2} = 2 + \frac{2}{s+1} + \frac{-2}{s+2}$$

thus,

$$f(t) = \mathscr{L}^{-1}[F(s)] = 2\delta(t) + 2e^{-t}u(t) - 2e^{-2t}u(t)$$

(2) $F(s)$ 的极点为共轭复数根(complex conjugate roots)

Suppose that there are a complex conjugate pair of poles ($p_{1,2} = -\alpha \pm j\beta$) in the denominator polynomial. $F(s)$ can be represented as

$$F(s) = \frac{N(s)}{(s + a - j\beta)(s + a + j\beta)D_1(s)} \quad (n > m)$$

the corresponding partial fraction expansion can take following forms

$$F(s) = \frac{k_1}{s + a - j\beta} + \frac{k_2}{s + a + j\beta} + \frac{k_3}{s - p_3} + \cdots + \frac{k_n}{s - p_n} \quad (4-46)$$

here, $p_k (3 \leq k \leq n)$ are real distinct roots and

$$k_1 = (s + a - j\beta)F(s)\big|_{s=-a+j\beta} = \frac{N(-a + j\beta)}{2j\beta D_1(-a + j\beta)}$$

$$k_2 = (s + a + j\beta)F(s)\big|_{s=-a-j\beta} = \frac{N(-a - j\beta)}{-2j\beta D_1(-a - j\beta)}$$

let $k_1 = A+jB$ and $k_2 = A-jB = k_1^*$, it is easy to find the inverse Laplace transform for the first two terms of $F(s)$. That is,

$$\mathscr{L}^{-1}\left[\frac{k_1}{s+a-j\beta} + \frac{k_1^*}{s+a+j\beta}\right] = k_1 e^{-at} e^{j\beta t} + k_1^* e^{-at} e^{-j\beta t}$$
$$= [(A+jB)e^{j\beta t} + (A-jB)e^{-j\beta t}]e^{-at}$$
$$= 2e^{-at}[A\cos\beta t - B\sin\beta t] \quad (4-47)$$

Example 4-18: Find the inverse transform of $F(s)$.

$$F(s) = \frac{s^2+3}{(s+2)(s^2+2s+5)}$$

Solution: We note that the denominator polynomial can be easily factored.

$$F(s) = \frac{s^2+3}{(s+2)(s+1-2j)(s+1+2j)}$$
$$= \frac{k_1}{s+1-2j} + \frac{k_1^*}{s+1-2j} + \frac{k_3}{s+2}$$

$$k_1 = \frac{s^2+3}{(s+2)(s+1+2j)}\bigg|_{s=-1+2j} = -\frac{1}{5} + j\frac{2}{5}$$

that is, $A = -\frac{1}{5}$ and $B = \frac{2}{5}$.

$$k_3 = \frac{s^2+3}{s^2+2s+5}\bigg|_{s=-2} = \frac{7}{5}$$

according to the formula (4-44) and (4-47), the inverse transform of the $F(s)$ is

$$f(t) = \frac{7}{5}e^{-2t}u(t) - 2e^{-t}\left(\frac{1}{5}\cos 2t + \frac{2}{5}\sin 2t\right)u(t)$$

Example 4-19: Find the inverse of $F(s) = \frac{s}{s^2+2s+5}$ by using some commonly Laplace transform pairs.

Solution: $F(s)$ can be expanded as

$$F(s) = \frac{s}{(s+1)^2+2^2} = \frac{s+1-1}{(s+1)^2+2^2} = \frac{s+1}{(s+1)+2^2} - \frac{1}{2}\frac{2}{(s+1)^2+2^2}$$

according to the Eq. (4-16) and Eq. (4-17)

$$e^{-at}\cos\omega t u(t) \xleftrightarrow{\mathscr{L}} \frac{s+a}{(s+a)^2+\omega^2}, \quad \sigma > -a$$

$$e^{-at}\sin\omega t u(t) \xleftrightarrow{\mathscr{L}} \frac{\omega}{(s+a)^2+\omega}, \quad \sigma > -a$$

we can obtain the inverse transform is

$$f(t) = \mathscr{L}^{-1}\left[\frac{s+1}{(s+1)+2^2} - \frac{1}{2}\frac{2}{(s+1)^2+2^2}\right] = \left[e^{-t}\cos 2t - \frac{1}{2}e^{-t}\sin 2t\right]u(t)$$

(3) $F(s)$ 具有多重极点(repeat poles)

若 $F(s)$ 在 $s=p_1$ 处具有 r 重根及 $n-r$ 个实数单根,则可将 $F(s)$ 展开为:

$$F(s) = \frac{k_{11}}{s-p_1} + \frac{k_{12}}{(s-p_1)^2} + \cdots + \frac{k_{1k}}{(s-p_1)^k} + \cdots + \frac{k_2}{s-p_2} + \cdots + \frac{k_{n-r}}{s-p_{n-r}} \quad (n>m) \quad (4-48)$$

式中 $k_{1k}(k=1, 2, \cdots, r)$ 为 r 重极点对应的待定系数，可采用下面方法计算出来。

将式(4-45)两端同乘以 $(s-p_1)^r$，有

$$(s-p_1)^r F(s) = k_{11}(s-p_1)^{r-1} + \cdots + k_{1k}(s-p_1)^{r-k} + \cdots + \frac{k_2}{s-p_2}(s-p_1)^r + \cdots + \frac{k_{n-r}}{s-p_{n-r}}(s-p_1)^r \tag{4-49}$$

令 $s=p_1$，代入式(4-49)，可得

$$k_{1r} = [(s-p_1)^r F(s)]_{s=p_1} \tag{4-50}$$

对式(4-49)求一阶导数并令 $s=p_1$，可得 $k_{1,r-1} = \dfrac{\mathrm{d}}{\mathrm{d}s}[(s-p_1)^r F(s)]_{s=p_1}$

同理，对式(4-49)求 $(r-k)$ 阶导数，并令 $s=p_1$，可得

$$\frac{\mathrm{d}^{r-k}}{\mathrm{d}s^{r-k}}[(s-p_1)^r F(s)]_{s=p_1} = (r-k)! \, k_{1k}$$

即

$$k_{1k} = \frac{1}{(r-k)!}\frac{\mathrm{d}^{r-k}}{\mathrm{d}s^{r-k}}[(s-p_1)^r F(s)]_{s=p_1} \qquad k=1, 2, \cdots, r \tag{4-51}$$

由公式(4-19)和拉氏变换的时移性质，可以求出 $\dfrac{k_{1k}}{(s-p_1)^k}$ 项的拉氏反变换

$$\mathscr{L}^{-1}\left[\frac{k_{1k}}{(s-p_1)^k}\right] = \frac{k_{1k}}{(k-1)!}\mathscr{L}^{-1}\left[\frac{(k-1)!}{(s-p_1)^k}\right] = \frac{k_{1k}}{(k-1)!}t^{k-1}\mathrm{e}^{p_1 t}\mathrm{u}(t) \quad (k=1, 2, \cdots, r) \tag{4-52}$$

Example 4-20: Find the inverse Laplace transform of $F(s) = \dfrac{s+2}{s(s+3)(s+1)^2}$

Solution: $F(s)$ can be expanded as

$$F(s) = \frac{k_{11}}{s+1} + \frac{k_{12}}{(s+1)^2} + \frac{k_3}{s} + \frac{k_4}{s+3}$$

the inverse transform $f(t)$ is

$$f(t) = \left[\frac{k_{11}}{(1-1)!}t^{1-1}\mathrm{e}^{-t} + \frac{k_{12}}{(2-1)!}t^{2-1}\mathrm{e}^{-t} + k_3 + k_4\mathrm{e}^{-3t}\right]\mathrm{u}(t)$$
$$= (k_{11}\mathrm{e}^{-t} + k_{12}t\mathrm{e}^{-t} + k_3 + k_4\mathrm{e}^{-3t})\mathrm{u}(t)$$

We can evaluate the coefficient k_{11}, k_{12} and k_3, k_4 by Eq. (4-44) and Eq. (4-51) respectively. Therefore, we get

$$k_{11} = \frac{\mathrm{d}}{\mathrm{d}s}[(s+1)^2 F(s)]_{s=-1}$$
$$= \frac{\mathrm{d}}{\mathrm{d}s}\left[\frac{s+2}{s^2+3s}\right]_{s=-1}$$
$$= \frac{s(s+3)-(s+2)(2s+3)}{(s^2+3s)^2}\bigg|_{s=-1} = -\frac{3}{4}$$

$$k_{12} = (s+1)^2 \cdot \frac{s+2}{(s^2+3s)^2}\bigg|_{s=-1} = \frac{s+2}{s^2+3s}\bigg|_{s=-1} = -\frac{1}{2}$$

$$k_3 = \frac{2}{3} \qquad k_4 = \frac{1}{12}$$

so that

$$f(t) = \left[\frac{-3}{4}e^{-t} - \frac{1}{2}te^{-t} + \frac{2}{3} + \frac{1}{12}e^{-3t}\right]u(t)$$

Example 4-21: Find the inverse of the bilateral Laplace transform of $F_b(s)$, $F_b(s)$ is given as $F_b(s) = \dfrac{2s+3}{s^2+3s+2}$ (1) $\sigma>-1$ (2) $\sigma<-2$ (3) $-2<\sigma<-1$.

Solution: $F_b(s) = \dfrac{2s+3}{s^2+3s+2} = \dfrac{1}{s+1} + \dfrac{1}{s+2}$

From Eq. (4-7) and Eq. (4-8) we see that there are two possible inverse transforms for the form $\dfrac{1}{s+a}$, which depend on whether the ROC is to the left or the right of the pole. So the inverse transform of the individual terms in $F_b(s)$ can be obtained as follow:

① $\sigma>-1$, $\qquad f(t) = (e^{-t} + e^{-2t})u(t)$
② $\sigma<-2$, $\qquad f(t) = (-e^{-t} + -e^{-2t})u(-t)$
③ $-2<\sigma<-1$, $\qquad f(t) = -e^{-t}u(-t) + e^{-2t}u(t)$

4.4.2 留数法(Residue Method)

由公式(4-5)求拉普拉斯反变换$f(t)$，就是计算下式的积分

$$f(t) = \frac{1}{2\pi j}\int_{\sigma-j\infty}^{\sigma+j\infty} F(s)e^{st}dt$$

根据复变函数理论，上式积分可以用留数定理来完成，即在$F(s)$的收敛域内，选择一条平行于$j\omega$轴的直线为积分路径，并在该直线两侧构造两个半径趋于无穷大的半圆形成两个闭合的积分路径，求出积分路径内被积函数$F(s)e^{st}$全部极点的留数和，便得到了$F(s)$拉普拉斯反变换。

根据留数定理及约当辅助定理，上面的积分可由下式求得:

$$\mathscr{L}^{-1}[F(s)] = \sum_{p_i} \operatorname{Re} s[F(s)e^{st}]$$

也就是

$$f(t) = \frac{1}{2\pi j}\oint_c F(s)e^{st}dt = \sum_{i=1}^{n} \operatorname{Re} s[F(s)e^{st}; p_i] \tag{4-53}$$

在计算式(4-53)的留数时，若极点p_i为$F(s)e^{st}$的单极点，其留数为

$$\operatorname{Re} s[p_i] = [(s-p_i)F(s)e^{st}]|_{s=p_i} \tag{4-54}$$

若极点p_i为$F(s)e^{st}$的r阶重极点，其留数为

$$\operatorname{Re} s[p_i] = \frac{1}{(r-1)!}\left[\frac{d^{r-1}}{ds^{r-1}}(s-p_i)^r F(s)e^{st}\right]_{s-p_i} \tag{4-55}$$

Example 4-22: Find the inverse Laplace transform of $F(s)$ by using the residue method.

$$F(s) = \frac{s+2}{s(s+3)(s+1)^2}$$

Solution: $p_1 = 0$, $p_2 = -3$ are single poles and $p_{3,4} = -1$ is pole that repeat two times. Thus,

$$\operatorname{Re} s[0] = \frac{s+2}{(s+3)(s+1)^2}e^{st}\Big|_{s=0} = \frac{2}{3}$$

$$\operatorname{Re} s[-3] = \frac{s+2}{s(s+1)^2} e^{st} \Big|_{s=-3} = \frac{1}{12} e^{-3t}$$

$$\operatorname{Re} s[-1] = \frac{d}{ds}\left[\frac{s+2}{s(s+3)} e^{st}\right]_{s=-1} = -\frac{3}{4} e^{-t} - \frac{1}{2} t e^{-t}$$

so that

$$f(t) = \sum_{i=1}^{3} \operatorname{Re} s[p_i] = \left[\frac{2}{3} + \frac{1}{12} e^{-3t} - \frac{1}{2}(t+\frac{3}{2}) e^{-t}\right] u(t)$$

可见，与例 4-20 计算结果相同。

4.5 连续 LTI 系统的复频域分析（Analysis of LTI Continuous-time System in Complex Frequency Domain）

4.5.1 常系数微分方程的复频域求解（Complex Frequency Domain Analysis of the Constant-coefficient Differential Equation）

拉普拉斯变换是求解常系数线性微分方程的有效方法，通过拉氏变换可以将时域微分方程变换为 s 域的代数方程，使得求解方便，而且系统的初始状态包含在象函数的代数方程中，求解代数方程可同时得出系统的全响应、零状态响应和零输入响应。

Now consider the LTI continuous-time system given by the second-order differential equation $\frac{d^2 r(t)}{dt^2} + a_1 \frac{dr(t)}{dt} + a_0 r(t) = b_1 \frac{de(t)}{dt} + b_0 e(t)$, the initial conditions of the system are $r(0^-)$, $r'(0^-)$, where a_1, a_0, $b_1 b_0$ are real numbers.

Taking the Laplace transform for both sides of the differential equation, we get

$$s^2 R(s) - s r(0^-) - r'(0^-) + a_1 [s R(s) - r(0^-)] + a_0 R(s) = b_1 s E(s) + b_0 E(s)$$

$$R(s) = \frac{r'(0^-) + s r(0^-) + a_1 r(0^-)}{s^2 + a_1 s + a_0} + \frac{b_1 s + b_0}{s^2 + a_1 s + a_0} E(s) \tag{4-56}$$

here we can find

$$R_{zi}(s) = \frac{r'(0^-) + s r(0^-) + a_1 r(0^-)}{s^2 + a_1 s + a_0}$$

$$R_{zs}(s) = \frac{b_1 s + b_0}{s^2 + a_1 s + a_0} E(s)$$

$$H(s) = \frac{b_1 s + b_0}{s^2 + a_1 s + a_0}$$

Where the $R_{zi}(s)$, $R_{zs}(s)$, $R(s)$ are the Laplace transform of the zero input response $r_{zi}(t)$, the zero state response $r_{zs}(t)$ and the complete response $r(t)$ of the system respectively.

Example 4-23: An LTI system described by the differential equation

$$\frac{d^2 r(t)}{dt^2} + 4 \frac{d}{dt} r(t) + 3 r(t) = \frac{de(t)}{dt} + 4 e(t)$$

the input signal $e(t) = u(t)$, and the initial conditions are $r(0^-) = 1$, $r'(0^-) = 2$. Find $r(t)$,

$r_{zi}(t)$, $r_{zs}(t)$ with Laplace Transform method.

Solution: Taking the unilateral Laplace transform for the differential equation on both sides, we get

$$[s^2R(s)-sr(0^-)-r'(0^-)]+4[sR(s)-r(0^-)]+3R(s)=sE(s)+4E(s)$$

$$(s^2+4s+3)R(s)=(s+4)E(s)+[sr(0^-)+r'(0^-)+4r(0^-)]$$

$$R(s)=\frac{(s+4)E(s)}{(s+1)(s+3)}+\frac{sr(0^-)+r'(0^-)+4r(0^-)}{(s+1)(s+3)}$$

$$=R_{zs}(s)+R_{zi}(s)$$

Substituting the initial conditions $r(0^-)=1$, $r'(0^-)=2$ and the Laplace transform of input signal $E(s)=\mathscr{L}[u(t)]=\frac{1}{s}$ yields

$$R(s)=\frac{(s+4)}{(s+1)(s+3)}\frac{1}{s}+\frac{s+6}{(s+1)(s+3)}=R_{zs}(s)+R_{zi}(s)$$

so that, the zero input response

$$r_{zi}(t)=\mathscr{L}^{-1}[R_{zi}(s)]=\mathscr{L}^{-1}\left[\frac{s+6}{(s+1)(s+3)}\right]=\mathscr{L}^{-1}\left[\frac{\frac{5}{2}}{s+1}+\frac{-\frac{3}{2}}{s+3}\right]=\left(\frac{5}{2}e^{-t}-\frac{3}{2}e^{-3t}\right)u(t)$$

the zero state response $r_{zs}(t)=\mathscr{L}^{-1}[R_{zs}(s)]$

$$R_{zs}(s)=\frac{(s+4)}{(s+1)(s+3)}\frac{1}{s}=\frac{K_1}{s}+\frac{K_2}{s+1}+\frac{K_3}{s+3}$$

$$K_1=sR_{zs}(s)|_{S=0}=\frac{s+4}{(s+1)(s+3)}\bigg|_{S=0}=\frac{4}{3}$$

$$K_2=(s+1)R_{zs}(s)|_{S=-1}=\frac{s+4}{s(s+3)}\bigg|_{S=-1}=-\frac{3}{2}$$

$$K_3=(s+3)R_{zs}(s)|_{S=-3}=\frac{-3+4}{-3\times(-2)}=\frac{1}{6}$$

$$r_{zs}(t)=\mathscr{L}^{-1}[R_{zs}(s)]=\left(\frac{4}{3}-\frac{3}{2}e^{-t}+\frac{1}{6}e^{-3t}\right)u(t)$$

so that, the complete response $r(t)=r_{zi}(t)+r_{zs}(t)=\left(\frac{4}{3}+e^{-t}-\frac{4}{3}e^{-3t}\right)u(t)$

由例4-23可见，单边拉普拉斯变换可以直接求解非零初始条件下LTI系统的微分方程，不管输入信号是否存在跳跃间断点。与傅里叶变换只能求解系统的零状态响应不同，拉普拉斯变换能够求出系统的零输入响应和零状态响应，因此也就能够给出系统的完全解。

4.5.2 电路的 s 域模型(Laplace Transform Analysis for Circuits Systems)

在分析具体电路时，利用电路的 s 域模型可以跳过列写系统微分方程这一步骤，直接得到象函数的代数方程，然后求解。

Resistor, inductor and capacitor are the basic elements in the circuits systems. We first establish the mode of the resistors, inductors and the capacitors in the complex frequency domain, then using these basic concepts to develop the method of analysis. The resistor, inductor and capacitor are defined in term of the voltage-current relationships：

$$v_R(t) = Ri_R(t) \tag{4-57a}$$

$$v_L(t) = L\frac{di_L(t)}{dt} \tag{4-57b}$$

$$v_C(t) = \frac{1}{C}\int_{0^-}^{t} i_C(\tau)d\tau + u_c(0^-) \tag{4-57c}$$

对式(4-57a)~式(4-57c)进行拉氏变换,可得

$$V_R(s) = RI_R(s) \tag{4-58a}$$

$$V_L(s) = LsI(s) - Li_L(0^-) \tag{4-58b}$$

$$V_C(s) = \frac{1}{sC}I_C(s) + \frac{1}{s}v_c(0^-) \tag{4-58c}$$

由式(4-58a)~式(4-58c)可以对 R,L,C 元件构成一种串联形式的 s 域模型。若将式(4-57a)至式(4-57c)进行拉氏变换并对电流求解,可得

$$I_R(s) = \frac{V_R(s)}{R} \tag{4-59a}$$

$$I_L(s) = \frac{1}{sL}V_L(s) + \frac{1}{s}i_L(0^-) \tag{4-59b}$$

$$I_C(s) = C[sV_C(s) - v_C(0^-)] \tag{4-59c}$$

利用式(4-59a)~式(4-59c)可以对 R,L,C 元件构成一种并联形式的 s 域模型。上述元件的 s 域模型如图 4-5~图 4-7 所示。

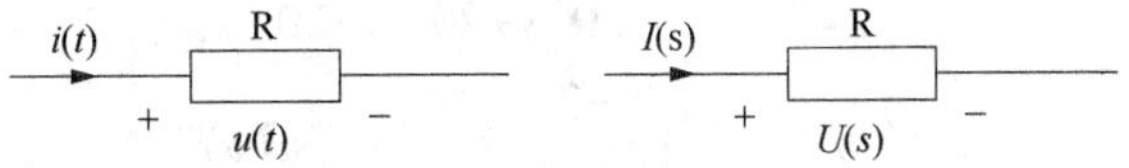

图 4-5 电阻的 s 域模型

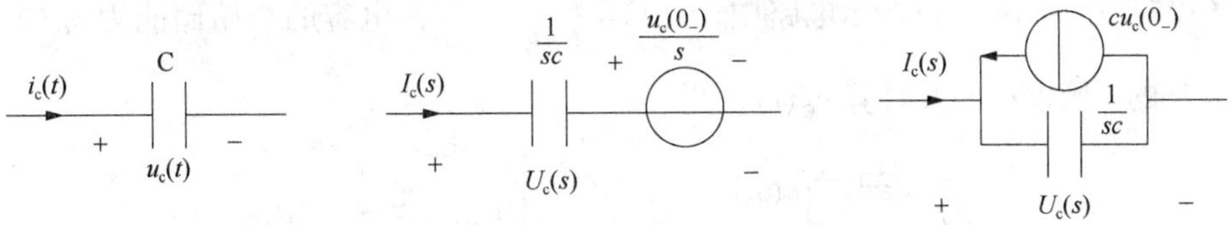

图 4-6 电容的 s 域模型

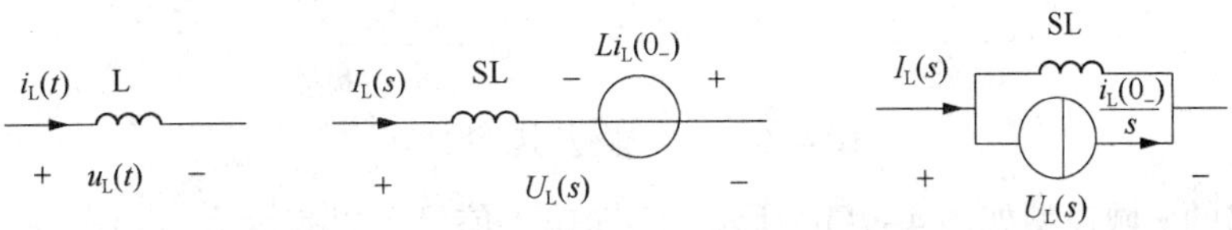

图 4-7 电感的 s 域模型

将电路中各元件用相应的 s 域模型代替,就得到了电路的 s 域模型。对此模型利用 KVL 和 KCL 分析,可列出有关待求响应象函数的代数方程,解代数方程求出响应的象函数,再通过拉氏反变换获得响应的时域解。

Example 4-24: Using Laplace transform analysis method to determine the voltage $r(t)$ in the circuit of Figure 4-8 for an input voltage $e(t) = 3e^{-10t}u(t)$. The voltage across the capacitor at time $t = 0^-$ is 5V.

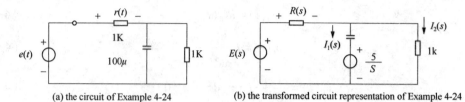

(a) the circuit of Example 4-24　　(b) the transformed circuit representation of Example 4-24

Figure 4-8

Solution: The transformed circuit is drawn in Figure 4-8(b), with symbol $I_1(s)$ and $I_2(s)$ representing the current through each branch. Using Kirchhoff's law, we write the following equations to describe the circuit.

$$R(s) = 1000[I_1(s)+I_2(s)]$$

$$E(s) = R(s) + \frac{1}{10^{-4}s}I_1(s) + \frac{5}{s}$$

$$E(s) = R(s) + 1000 I_2(s)$$

combining these three equations to eliminate $I_1(s)$ and $I_2(s)$, yields

$$R(s) = E(s) \cdot \frac{s+10}{s+20} - \frac{5}{s+20}$$

using $E(s) = \frac{3}{s+10}$, we obtain

$$R(s) = \frac{3}{s+20} - \frac{5}{s+20} = \frac{-2}{s+20}$$

thus

$$r(t) = \mathscr{L}^{-1}\left[\frac{-2}{s+20}\right] = -2e^{-20t}u(t)$$

例 4-25：在图 4-9(a)所示电路中，$e(t) = \begin{cases} -E & t<0 \\ E & t>0 \end{cases}$，电容的初始储能为 $v_C(0_-) = -E$，画出该电路的 s 域模型，并计算 $v_C(t)$。

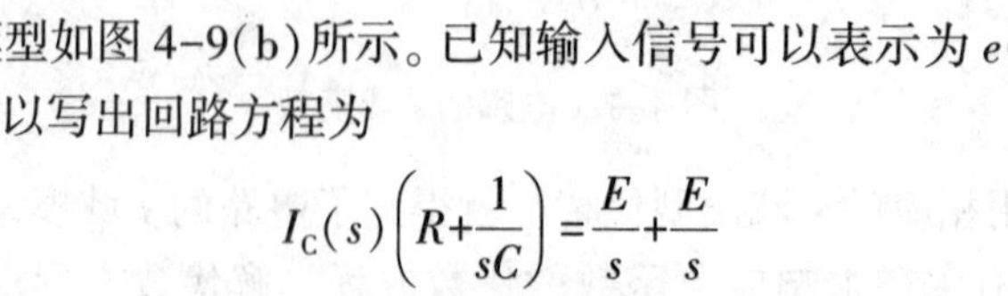

(a) RC 电路　　(b) s 域模型

图 4-9　RC 电路及其 s 域模型

解：电路的 s 域模型如图 4-9(b)所示。已知输入信号可以表示为 $e(t) = -Eu(-t)+Eu(t)$，根据电路的 s 域模型可以写出回路方程为

$$I_C(s)\left(R+\frac{1}{sC}\right) = \frac{E}{s} + \frac{E}{s}$$

求出回路电流

$$I_C(s) = \frac{2E}{s\left(R+\frac{1}{sC}\right)}$$

求出电容两端的电压 $V_C(s)$

$$V_C(s) = I_C(s) \cdot \frac{1}{sC} + \frac{-E}{s}$$

经整理得

$$V_C(s) = \frac{E}{s} - \frac{2E}{s + \frac{1}{RC}}$$

对 $V_C(s)$ 求拉氏反变换得

$$v_C(t) = E(1 - 2\mathrm{e}^{-\frac{t}{RC}}) \quad (t \geqslant 0)$$

响应时域波形图如图 4-10 所示。

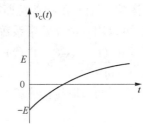

图 4-10 响应 $v_C(t)$ 的时域波形图

4.6 系统函数与系统特性（System Function and System Characteristic）

系统函数 $H(s)$ 是描述连续时间系统特性的重要特征参数，利用系统函数的零极点分布，可定性分析系统的时域特性、频率响应和稳定性等特性。

4.6.1 系统函数(传递函数)(System Function(Transfer Function))

If an LTI system is given by an input/output differential equation, we assume that the impulse response of the system is $h(t)$ and the zero state response of the system (i.e., the system is at rest) for arbitrary input $e(t)$ can be obtained by

$$r(t) = e(t) * h(t) \tag{4-60}$$

Taking the Laplace transform on both sides of Eq. (4-60) yield

$$R(s) = E(s)H(s) \tag{4-61}$$

Where, $R(s)$, $E(s)$, $H(s)$ are the Laplace transform of the output $r(t)$, the input $e(t)$ and the impulse response $h(t)$ respectively.

LTI 连续时间系统的系统函数(system function)定义为系统零状态响应的拉氏变换 $R(s)$ 与系统激励的拉氏变换 $E(s)$ 之比，也称传递函数(transfer function)，即

$$H(s) = \frac{R(s)}{E(s)} \tag{4-62}$$

Example 4-26：Consider a system with the differential equation

$$\frac{\mathrm{d}^2 r(t)}{\mathrm{d}t^2} + 3\frac{\mathrm{d}r(t)}{\mathrm{d}t} + 2r(t) = 2\frac{\mathrm{d}e(t)}{\mathrm{d}t} + 3e(t)$$

find the system function $H(s)$ and the impulse response $h(t)$.

Solution: Taking the Laplace transform on both sides of the differential equation yields

$$(s^2+3s+2)R(s) = (2s+3)E(s)$$

by definition of the system function, we get

$$H(s) = \frac{R(s)}{E(s)} = \frac{2s+3}{s^2+3s+2} = \frac{(s+1)+(s+2)}{(s+1)(s+2)}$$

$$= \frac{1}{s+1} + \frac{1}{s+2}$$

therefore, the impulse response

$$h(t) = \mathscr{L}^{-1}[H(s)] = (e^{-t}+e^{-2t})u(t)$$

可见，系统函数 $H(s)$ 是系统冲激响应 $h(t)$ 的拉普拉斯变换。

对于一个线性时不变连续系统，可以用一个 n 阶常系数微分方程表示为

$$\sum_{i=0}^{n} a_i \frac{d^i r(t)}{dt^i} = \sum_{j=0}^{m} b_j \frac{d^j e(t)}{dt^j}$$

设系统的初始状态为零，对微分方程两边取拉氏变换并利用拉氏变换的时域微分性质，可得

$$\left\{\sum_{i=0}^{n} a_i s^i\right\} R(s) = \left\{\sum_{j=0}^{m} b_j s^j\right\} E(s)$$

根据系统函数的定义，则有

$$H(s) = \frac{\sum_{j=0}^{m} b_j s^j}{\sum_{i=0}^{n} a_i s^i} \tag{4-63}$$

可见，系统函数与系统微分方程有一一对应的关系。系统函数可以由系统的微分方程获得，同理，知道系统函数也可以写出系统的微分方程。

Example 4-27: The impulse response of an LTI causal system is

$$h(t) = e^{-2t}u(t) - \delta(t)$$

Find: ① The system function of the system. ② The differential equation of the system.

Solution: ① Because the system function $H(s)$ and impulse response $h(t)$ is a Laplace transform pair

$$H(s) = \mathscr{L}[h(t)] = \mathscr{L}[e^{-2t}u(t) - \delta(t)] = \frac{1}{s+2} - 1 = \frac{-s-1}{s+2} \quad \sigma > -2$$

② From the definition of $H(s)$, we have $R(s) = H(s)E(s)$

$$R(s)(s+2) = E(s)(-s-1)$$

Taking the inverse Laplace transform on both sides of the above equation, we get the differential equation of the system

$$\frac{dr(t)}{dt} + 2r(t) = -\frac{de(t)}{dt} - e(t)$$

4.6.2 系统函数的零极点分布(Zeros-poles Distribution of the System Function)

对于一个 n 阶常系数微分方程，两边取拉氏变换并利用拉氏变换的时域微分性质，可得

$$\{\sum_{i=0}^{n} a_i s^i\} R(s) = \{\sum_{j=0}^{m} b_j s^j\} E(s)$$

根据系统函数的定义，则有

$$H(s) = \frac{\sum_{j=0}^{m} b_j s^j}{\sum_{i=0}^{n} a_i s^i} = \frac{b_m s^m + b_{m-1} s^{m-1} + \cdots + b_1 s + b_0}{a_n s^n + a_{n-1} s^{n-1} + \cdots + a_1 s + a_0} \quad (4-64)$$

系统函数 $H(s)$ 分母多项式 $D(s) = \sum_{i=0}^{n} a_i s^i = 0$ 的根，称为系统函数 $H(s)$ 的极点；$H(s)$ 分子多项式 $N(s) = \sum_{j=0}^{m} b_j s^j = 0$ 的根，称为系统函数 $H(s)$ 的零点。

由此，上式可改写为

$$H(s) = \frac{N(s)}{D(s)} = \frac{b_m s^m + b_{m-1} s^{m-1} + \cdots + b_1 s + b_0}{a_n s^n + a_{n-1} s^{n-1} + \cdots + a_1 s + a_0}$$

$$= H_0 \cdot \frac{(s-z_1)(s-z_2)\cdots(s-z_m)}{(s-p_1)(s-p_2)\cdots(-p_n)}$$

$$= H_0 \cdot \frac{\prod_{j=1}^{m}(s-z_j)}{\prod_{i=1}^{n}(s-p_i)} \quad (4-65)$$

式中，$z_j(j=1, 2\cdots m)$ 为系统函数的零点，$p_i(i=1, 2\cdots n)$ 为系统函数的极点。$H_0 = \frac{b_m}{a_n}$ 是系统的增益。

Given all the poles and zeros of the system and gain factor completely, we can determine the system function $H(s)$ and thus also offers another description of an LTI system.

极点和零点的数值有三种情况：实数，纯虚数和复数。由于系统函数 $N(s)$ 和 $D(s)$ 的系数都是实数，因此，零极点中若有虚数或复数时必共轭成对出现。在 s 平面上，画出 $H(s)$ 的零极点图时，极点用"×"表示，零点用"○"表示。

例 4-28：已知某系统的系统函数为 $H(s) = \frac{s^3 - 4s^2 + 5s}{s^4 + 4s^3 + 5s^2 + 4s + 4}$，求系统的零、极点，并画出系统的零极点分布图。

解：系统函数 $H(s)$ 可表示为

$$H(s) = \frac{s(s-2-j)(s-2+j)}{(s+2)^2(s-j)(s+j)}$$

系统的零极点分别为

$$z_1 = 0, \quad z_2 = 2+j, \quad z_3 = 2-j$$

$$p_{1,2} = -2, \quad p_3 = j, \quad p_4 = -j$$

零极点分布图如图 4-11 所示。

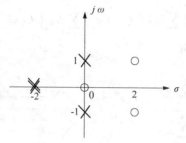

图 4-11 零极点分布图

4.6.3 系统函数的零极点分布与系统时域特性的关系(Relationships between the Impulse Response and Zeros-poles Location of the System Function)

在 s 域分析中，借助系统函数在 s 平面零点与极点分布的研究，可以直观地给出系统响应的许多规律。下面讨论由系统函数的零极点分布来确定系统的时域特性。

冲激响应 $h(t)$ 与系统函数 $H(s)$ 从时域和变换域两方面表征了同一系统的特性。从前面讨论可知冲激响应 $h(t)$ 与系统函数 $H(s)$ 是一对拉氏变换对，可以将 $H(s)$ 展开为部分分式，假设 $H(s)$ 为有理真分式，且 $H(s)$ 所有极点均为单极点，则 $H(s)$ 可以展开为：

$$H(s) = H_0 \cdot \frac{(s-z_1)(s-z_2)\cdots(s-z_m)}{(s-p_1)(s-p_2)\cdots(-p_n)} = \sum_{k=1}^{n} \frac{A_k}{s-p_k} \tag{4-66}$$

对每个分式进行拉氏反变换可得

$$h(t) = \mathscr{L}^{-1}[H(s)] = \sum_{k=1}^{n} A_k e^{p_k t} u(t) \tag{4-67}$$

We see that the form of the $h(t)$ completely depend on the poles of the system function $H(s)$. And the zeros and poles together determine the amplitude value of $h(t)$.

根据 $H(s)$ 的零极点在 s 平面的位置来确定系统的时域特性，下面讨论 $H(s)$ 的典型极点分布与冲激响应 $h(t)$ 波形的对应关系，如图 4-12 所示，具体分析如下。

若 $H(s)$ 具有一阶极点 $p_1 = -a$ 位于 s 平面的实轴上，即

$$H(s) = \frac{k}{s+a}, \quad \sigma > -a$$

① 若 $a=0$，极点在原点，$H(s) = \frac{k}{s}$，$\sigma > 0$。相应的冲激响应 $h(t)$ 为阶跃信号，$h(t) = ku(t)$。

That means if $a=0$, then the pole lies in the origin, the impulse response $h(t)$ is equal to the step response $ku(t)$.

② 若 $a>0$，极点为负实数，相应冲激响应 $h(t)$ 为指数衰减信号，即 $h(t) = ke^{-at}u(t)$。

If a is positive, the pole is in the left half of the s-plane. As t increases ($t \to \infty$), the time response $h(t)$ dies away to zero $[h(t) \to 0]$.

③ 若 $a<0$，极点为正实数，相应冲激响应 $h(t)$ 为指数增长信号，即 $h(t) = ke^{-at}u(t)$。

If a is negative, so that the pole is in the right half of the s-plane, then the time response increases $[h(t) \to \infty]$ without bound as t increases ($t \to \infty$).

若 $H(s)$ 具有一对共轭极点 $p_{1,2} = -a \pm j\omega_0$，即

$$H(s) = \frac{C}{(s+a)^2 + \omega_0^2}, \quad \sigma > -a$$

① 若 $a=0$，这对共轭极点位于 $j\omega$ 轴上，$H(s)=\dfrac{C}{s^2+\omega_0^2}$，$\sigma>0$。相应冲激响应 $h(t)$ 为等幅振荡的正弦信号，即 $h(t)=\dfrac{C}{\omega_0}\sin\omega_0 t u(t)$。

In this case, the impulse response is a sinusoidal signal with the same amplitude oscillation.

② 若 $a>0$，$H(s)$ 具有位于 s 平面左半平面上的一对共轭极点，$p_{1,2}=-a\pm j\omega_0$，冲激响应 $h(t)$ 对应于衰减振荡，即 $h(t)=\dfrac{C}{\omega_0}e^{-at}\sin\omega_0 t u(t)$。

In this case, there is exponential damping, and so the response die away as t increases.

③ 若 $a<0$，$H(s)$ 具有位于 s 平面右半平面上的一对共轭极点，$p_{1,2}=-a\pm j\omega_0$，冲激响应 $h(t)$ 对应于增幅振荡，即 $h(t)=\dfrac{C}{\omega_0}e^{|a|t}\sin\omega_0 t u(t)$。

In this case, there is exponential increasing, and so the response increases [$h(t)\to\infty$] without bound as t increases ($t\to\infty$).

We see that left-plane poles have time functions that die away as t increases and right plane poles correspond to time functions that increase without bound as t increases.

若 $H(s)$ 具有多重极点，这里以二阶极点为例讨论。

① $H(s)=\dfrac{1}{s^2}$，极点在原点，则相应冲激响应 $h(t)=tu(t)$，$t\to\infty$，$h(t)\to\infty$。

② $H(s)=\dfrac{2\omega_0 s}{(s^2+\omega_0^2)^2}$，极点在虚轴上，则相应冲激响应 $h(t)=t\sin\omega_0 t u(t)$。$t\to\infty$，$h(t)$ 增幅振荡。

③ $H(s)=\dfrac{1}{(s+a)^2}$ ($a>0$)，极点在实轴上，则相应冲激响应 $h(t)=te^{-at}u(t)$，$a>0$，$t\to\infty$，$h(t)\to 0$。

综上所述，若系统函数 $H(s)$ 的极点位于 s 平面左半平面，则冲激响应 $h(t)$ 的波形呈衰减形式变化；若 $H(s)$ 的极点位于 s 平面右半平面，则 $h(t)$ 的波形呈增幅变化；当一阶极点位于虚轴时，冲激响应 $h(t)$ 成阶跃信号或等幅振荡变化，但是当二阶极点位于虚轴时，相应 $h(t)$ 呈增幅变化。

可见 $H(s)$ 的零点分布仅影响冲激函数 $h(t)$ 的振幅和相位，对冲激函数 $h(t)$ 的波形形式不起作用，而 $H(s)$ 的极点分布决定冲激函数 $h(t)$ 的函数形式，如图 4-12 所示。

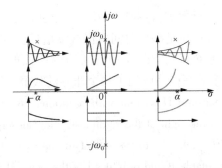

图 4-12　典型极点分布与冲激响应 $h(t)$ 波形的对应关系

4.6.4 系统函数与系统的稳定性(System Function and Stability)

稳定性是系统自身的性质之一，系统是否稳定与激励信号的情况无关。冲激响应 $h(t)$ 和系统函数 $H(s)$ 集中表征了系统的特性，当然它们也反映了系统是否稳定，下面从这两个方面确定系统的稳定性。

如果输入有界时(bounded input)能产生有界(bounded output)输出的系统，则该系统称为稳定系统，这一稳定性准则称为 BIBO 稳定性准则。它适用于一般系统，可以是线性系统也可以是非线性系统，可以是非时变系统也可以是时变系统，可以是因果系统也可以是非因果系统。

对于 LTI 连续时间系统，从时域判断其是否为 BIBO 稳定系统，只要判断该系统的冲激响应 $h(t)$ 是否绝对可积，即

$$\int_{-\infty}^{\infty} |h(t)| dt < \infty \tag{4-68}$$

则系统为稳定系统，式(4-68)是连续系统稳定的充要条件，对于 LTI 因果系统上式可改写为

$$\int_{0}^{\infty} |h(t)| dt < \infty \tag{4-69}$$

The ROC of $H(s)$ can also be raelated to the stability of a system. The stability of an LTI system is equivalent to its impulse response being absolutely integrable.

式(4-68)及式(4-69)是积分式，以此来判断系统的稳定性还是很麻烦的，对于因果系统，式(4-69)是否收敛，取决于 $h(t)$ 是否有界，因为 $h(t)$ 和 $H(s)$ 是一对拉氏变换对，所以也可以从系统函数 $H(s)$ 的极点在 s 平面的位置来判断系统的稳定性。

从稳定性角度考虑，因果系统可划分为稳定系统、不稳定系统和临界稳定系统三种情况。

① 稳定系统：如果系统函数 $H(s)$ 全部极点位于 s 平面的左半平面(不包含虚轴)，则可满足 $h(t)$ 有界，即

$$\lim_{t \to \infty} h(t) = 0 \tag{4-70}$$

也可以从系统函数 $H(s)$ 的收敛域判断，对于连续 LTI 系统，当系统函数 $H(s)$ 的收敛域包含 $j\omega$ 轴时，系统稳定。

An LTI system is stable if and only if the ROC of its system function $H(s)$ includes the entire $j\omega$-axis.

② 不稳定系统：如果系统函数 $H(s)$ 的极点位于 s 平面的右半平面，或虚轴上具有二阶以上极点，则系统冲激响应 $h(t)$ 在足够长时间后仍继续增长，系统是不稳定的。

③ 临界稳定系统：如果系统函数 $H(s)$ 的极点落于 s 平面的虚轴上，且只有一阶，则足够长时间以后，$h(t)$ 形成一个等幅振荡或趋于一个非零的数值，这时系统临界稳定。

Specifically, consider a causal LTI system with a rational system function $H(s)$. Since the system is causal, the ROC is to the right of the rightmost pole. Consequently, for this system to be stable, the tightmost pole of $H(s)$ must be to the left of the $j\omega$-axis. That is, a causal system with rational system function $H(s)$ is stable if and only if all of the poles of $H(s)$ lie in the left half of the s-plane—i. e. , all of the poles have negative real parts.

需要指出的是，利用上述方法判断系统的稳定性，必须知道系统函数全部极点的位置。当系统阶次较高时，求解 $H(s)$ 的极点也比较困难，利用一些其他判断方法如根轨迹法、罗斯判别法和图形判别法等也可以较为方便地判断系统的稳定性。

Example 4-29: A causal continuous-time LTI system has the following system function, determine whether the system is stable.

$$H(s) = \frac{s+3}{(s+1)(s+2)}$$

Solution: Because the system is a causal continuous-time LTI system, so we can see that all poles ($p_1 = -1$, $p_2 = -2$) of $H(s)$ are in the left half of the s-plane.

Therefore, the system is stable.

Example 4-30: A system has the transfer function

$$H(s) = \frac{4}{s+3} + \frac{1}{s-4}$$

① Please find the impulse response if assuming that the system is stable.
② Please find the impulse response if assuming that the system is causal.
③ Can this system be both stable and causal?

Solution: The system has two poles $p_1 = -3$ and $p_2 = 4$.

① If the system is stable, then the pole $p_1 = -3$ contributes a right-sided term to the impulse response, while the pole $p_2 = 4$ contributes a left-sided term. We thus have

$$h(t) = 4e^{-3t}u(t) - e^{4t}u(-t)$$

② If the system is causal, then both poles must contribute right-sided terms to the impulse response, and we have

$$h(t) = 4e^{-3t}u(t) + e^{4t}u(t)$$

③ Note that this system is not stable, since the term $e^{4t}u(t)$ is not absolutely integral. In fact, the system cannot be both stable and casual because the pole $p_2 = 4$ is in the right half of the s-plane.

4.6.5 系统函数的零极点分布与系统频率响应特性的关系(Relationships between the Frequency Response and Zeros-poles Location of System Function)

(1) 频率响应(frequency response)

频率响应是系统在虚指数信号(或正弦信号)激励下稳态响应随激励信号的频率变化情况。在控制工程领域中频率响应又称为频率特性，是指系统对不同频率的正弦信号的稳态响应特性。在系统理论中，正弦稳态响应是稳态响应中一个重要概念，下面简要介绍。

Assume that $H(s)$ is the transfer function of a stable and causal system and the poles of $H(s)$ are all distinct. The transformed output for an input $e(t) = E_m \sin\omega_0 t u(t)$ is

$$R(s) = E(s) \cdot H(s) = \frac{N(s)}{D(s)} \cdot \frac{E_m \omega_0}{s^2 + \omega_0^2}$$

$$= \frac{k_1}{s + j\omega_0} + \frac{k_2}{s - j\omega_0} + \sum_{i=3}^{n} \frac{k_i}{s - p_i}$$

where,
$$k_1 = (s+j\omega_0)R(s)|_{s=-j\omega} = \frac{E_m}{-2j}H(-j\omega_0)$$

$$k_2 = \frac{E_m}{2j} \cdot H(j\omega_0)$$

Because of $H(s)$ has distinct roots only in the left-hand half of the s-plane, the terms $\sum_{i=3}^{n}\frac{k}{s-p_i}$ represent transient terms in the time domain that die away as $t\to\infty$. Thus, the steady state output is

$$r(t) = \mathscr{L}^{-1}\left\{-\frac{E_m H(j\omega_0)}{2j(s+j\omega_0)} + \frac{E_m H(j\omega_0)}{2j(s-j\omega_0)}\right\}$$

$$= -E_m \frac{H(-j\omega_0)}{2j}e^{-j\omega_0 t} + E_m \frac{H(j\omega_0)}{2j}e^{j\omega_0 t} \tag{4-71}$$

$H(j\omega_0)$ can be written in polar form as
$$H(j\omega_0) = |H(j\omega_0)|e^{j\varphi_0} \quad \text{and} \quad H(-j\omega_0) = |H(j\omega_0)|e^{-j\varphi_0}$$

substituting $H(j\omega_0)$ and $H(-j\omega_0)$ into Eq. (4-71), we get

$$r_{zs}(t) = \frac{E_m}{2j}|H(j\omega_0)| \cdot [e^{j(\omega_0 t+\varphi_0)} - e^{-j(\omega_0 t-\varphi_0)}]$$

$$r_{zs}(t) = E_m|H(j\omega_0)| \cdot \sin(\omega_0 t + \varphi_0) \tag{4-72}$$

可见，对于稳定因果系统，在频率为 ω_0 的正弦信号激励下，系统的稳态响应仍为同频率的正弦信号，且幅度乘以系数 $|H(j\omega_0)|$，相位移动 φ_0，$|H(j\omega_0)|$ 和 φ_0 由系统函数在 $j\omega_0$ 处的取值决定。从式(4-72)也可看出，对于线性系统，它的正弦稳态响应是与激励同频率的正弦信号，即线性系统具有频率保持性。

当正弦信号的频率改变时，将变量 ω 代入到 $H(s)$ 之中，即可得到频率响应特性

$$H(s)|_{s=j\omega} = H(j\omega) = |H(j\omega)|e^{j\varphi(\omega)} \tag{4-73}$$

式中 $|H(j\omega)|$ 是系统的幅频响应特性，$\varphi(\omega)$ 是相频响应特性。使用中常常将它们绘制成频率特性曲线。

（2）频率特性曲线的几何画法（geometric drawing of frequency characteristic curve）

系统函数 $H(s)$ 在 s 平面的零极点分布与其频率特性有直接的关系，利用系统函数零极点分布可以方便地绘制频率响应特性曲线。

如前面所述，由系统的零极点分布可以写出零极点增益形式的系统函数 $H(s)$

$$H(s) = H_0 \frac{(s-z_1)(s-z_2)\cdots(s-z_m)}{(s-p_1)(s-p_2)\cdots(s-p_n)} \tag{4-74}$$

当 $H(s)$ 的极点全部位于 s 平面的左半平面时，令 $s=j\omega$，系统的频率响应 $H(j\omega)$ 可以由 $H(s)$ 求出，即

$$H(s)|_{s=j\omega} = H(j\omega) = |H(j\omega)|e^{j\varphi(\omega)} \tag{4-75}$$

也就是说，系统函数在 s 平面中令 s 沿虚轴移动变化，即可得到系统的频率响应 $H(j\omega)$。

The frequency response is obtained from the transfer function by substituting $j\omega$ for s, that is, by evaluating the transfer function along the $j\omega$-axis in the s-plane. Substituting $s=j\omega$ into Eq. (4-70), yield

$$H(j\omega) = H_0 \frac{(j\omega - z_1)(j\omega - z_2)\cdots(j\omega - z_m)}{(j\omega - p_1)(j\omega - p_2)\cdots(j\omega - p_n)}$$

$$= H_0 \frac{\prod_{i=1}^{m}(j\omega - z_i)}{\prod_{k=1}^{n}(j\omega - p_k)} \tag{4-76}$$

可以看出，频率响应取决于系统的零极点。根据零极点分布情况可以绘制出系统的频率响应曲线，包括幅频特性曲线$|H(j\omega)|$和相频特性曲线$\varphi(\omega)$。下面简要介绍由向量法绘制系统频响特性曲线。

复数值在复平面内可以用原点到复数坐标点的向量表示。式(4-76)中$j\omega$，p_i，z_j均为复数，因此分子中任一因子$(j\omega-z_j)$相当于由零点z_j引向虚轴上某点$j\omega$的一个向量，图4-13所表示的是由零点z_j与$j\omega$连接构成的向量，图中N_j，ψ_j分别表示向量的模和相角。

同理，对任意零点z_j、任意极点p_i，相应的向量都可以表示为

$$j\omega - z_j = |j\omega - z_j|e^{j\psi_j} = N_j e^{j\psi_j}(j=1,2,\cdots,m)$$

$$j\omega - p_i = |j\omega - p_i|e^{j\theta_i} = M_i e^{j\theta_i}(i=1,2,\cdots,n)$$

则式(4-76)可改写为

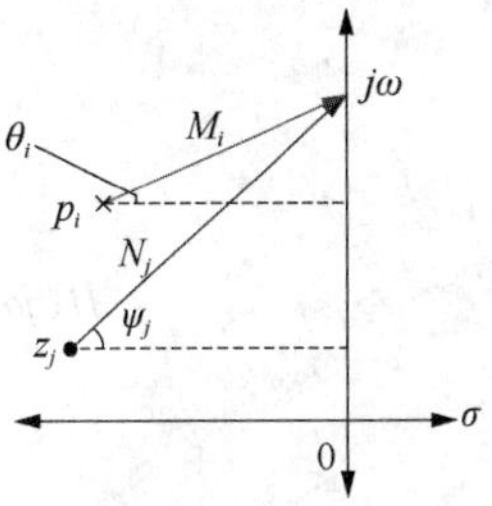

$$H(j\omega) = H_0 \frac{N_1 N_2 \cdots N_m}{M_1 M_2 \cdots M_n} e^{j(\psi_1+\psi_2+\cdots+\psi_m-\theta_1-\theta_2-\cdots-\theta_n)}$$

图4-13 $(j\omega-z_j)$和$(j\omega-p_i)$向量表示

$$= H_0 \frac{\prod_{j=1}^{m} N_j}{\prod_{i=1}^{n} M_i} \exp\left(\sum_{j=1}^{m}\psi_j - \sum_{i=1}^{n}\theta_i\right) \tag{4-77}$$

式中

$$|H(j\omega)| = H_0 \frac{\prod_{j=1}^{m} N_j}{\prod_{i=1}^{n} M_i} \tag{4-78}$$

$$\varphi(\omega) = \sum_{j=1}^{m}\psi_j - \sum_{i=1}^{n}\theta_i \tag{4-79}$$

当ω自$-\infty$沿着虚轴移动并逐渐趋于$+\infty$时，各零点向量$j\omega-z_j(j=1,2,\cdots,m)$和极点向量$j\omega-p_i(i=1,2,\cdots,n)$的模和相角都随之改变，于是得出系统的幅频特性曲线$|H(j\omega)|$和相频特性曲线$\varphi(\omega)$，如式(4-78)和式(4-79)。由前面章节可知系统的幅频响应偶对称、相频响应奇对称，绘制系统的频率响应曲线时仅绘出ω从$0\rightarrow+\infty$变化即可。

We can clearly see that the length of the vectors (N_j, M_i) are changing with ω, and in this way we may assess the contribution of each poles or zeros to the overall magnitude response. $\varphi(\omega)$ is the sum of the phase angles due to all the zeros, then subtracts the sum of the phase angle due to all the poles. These angles of a vector pointing from pole or zero to $j\omega$ in the s-plane. The angle of the vector is measured relative to the horizontal line through z_j or p_i. And we can see that the phase of the vectors are varying with ω. We may assess the contribution of each pole or zero to the overall phase frequency response in this way.

Example 4-31: Sketching the magnitude response and phase response of the low-pass filter as shown in Figure 4-14.

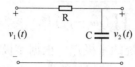

Figure 4-14 RC low-pass filter

Solution: The transfer function is given

$$H(s) = \frac{V_2(s)}{V_1(s)} = \frac{1}{RC} \cdot \left(\frac{1}{s+\frac{1}{RC}}\right) = \frac{\frac{1}{RC}}{s+\frac{1}{RC}}$$

$$H(j\omega) = H(s)\Big|_{s=j\omega} = \frac{\frac{1}{RC}}{j\omega+\frac{1}{RC}} = \frac{1}{RC}\frac{1}{M_1 e^{j\theta_1}} = \frac{V_2}{V_1}e^{j\varphi(\omega)}$$

where,

$$\frac{V_2}{V_1} = \frac{1}{RC}\frac{1}{M}, \quad \varphi = -\theta_1$$

the magnitude response is $|H(j\omega)| = \dfrac{\frac{1}{RC}}{\sqrt{\omega^2 + \left(\frac{1}{RC}\right)^2}} = \dfrac{V_2}{V_1}$

the phase response is $\varphi(\omega) = -\arctan\dfrac{\omega}{\frac{1}{RC}} = -\theta_1$

The low-pass filter has the pole at $p_1 = -\dfrac{1}{RC}$ as depicted in Figure 4-15(a). Hence the pole causes the magnitude response $|H(j\omega)|$ to increase as $\omega \to 0$ and decrease as $\omega \to \infty$, as depicted in Figure 4-15(b). Obviously, this system has the characteristics of low pass filter. The phase response is depicted in Figure 4-15(c).

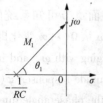
(a) the representation of vector $j\omega + \dfrac{1}{RC}$

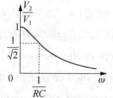
(b) the magnitude response of the low-pass filter

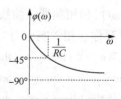
(c) the phase response of the low-pass filter

Figure 4-15 the vector representation and the magnitude and phase response of the low-pass filter

4.7 LTI 连续时间系统的模拟（Imitation of LTI Continuous-time Systems）

前面对系统的讨论都是将系统抽象为输入和输出的一种数学映射关系，如微分方程、系统函数等。系统的方框图(block diagram)模拟，是指用一些基本功能部件，经过合适的相互连接，实现微分方程或系统函数描述的系统功能。

Imitation of LTI continuous–time system (system block diagram) is given in terms of an interconnection of adders, multipliers, integrators to describe the behavior and properties of the system.

实现连续时间系统描述的基本部件有三种：加法器、乘法器和积分器，其数学模型及其方块图描述如下。

(1) 加法器(adder)

加法器的输出为两信号的和，即
$$f(t)=f_1(t)+f_2(t) \text{ 或 } f(t)=f_1(t)-f_2(t)$$

时域模型如图 4-16 所示。根据拉普拉斯变换的性质，可以方便地得到时域中各运算符号在 s 域中的模型(图 4-17)。

图 4-16　加法器时域模型　　　　图 4-17　加法器 s 域模型

(2) 乘法器(multiplier)

乘法器，也称数乘器模型，如下图所示，一般 a 是一个常数，表示对输入信号进行放大。乘法器的时域和 s 域模型分别如图 4-18 和图 4-19 所示。

图 4-18　乘法器时域模型　　　　图 4-19　乘法器 s 域模型

(3) 积分器(integrator)

积分器如下图所示，其输出为 $g(t)=\int_0^t f(x)\mathrm{d}x$。根据拉氏变换的积分性质，上式的 s 域表达式为 $G(s)=\dfrac{F(s)}{s}$。积分器的时域和 s 域模型分别如图 4-20 和图 4-21 所示。

图 4-20　积分器时域模型　　　　图 4-21　积分器 s 域模型

An LTI continuous–time system is sometimes specified by a block diagram consisting of interconnection of "blocks", with each block represented by a transfer function. The transfer functions for basic types of interconnections are direct interconnection, series interconnection, parallel interconnection and feedback connection.

下面介绍怎样利用加法器、乘法器和积分器这些基本部件来构建连续时间系统，使得系统的输入输出关系满足微分方程或系统函数。实现连续系统的方框图描述有直接、级联、并联和反馈连接四种实现形式。在此主要介绍直接实现形式。

Consider the causal continuous-time system described by a first-order differential equation
$$\frac{\mathrm{d}r(t)}{\mathrm{d}t}+a_0 r(t)=be(t)$$
in order to imitate this system, let us rewrite it as
$$\frac{\mathrm{d}r(t)}{\mathrm{d}t}=be(t)-a_0 r(t) \quad (4-80)$$
we can consider representing Eq. (4-80) as an interconnection of these basic elements, the block diagram is shown in Figure 4-22.

For the second-order differential equation
$$r''(t)+a_1 r'(t)+a_0 r(t)=e(t)$$
then
$$r''(t)=e(t)-a_0 r(t)-a_1 r'(t) \quad (4-81)$$

Similarly, we can represent this equation byinterconnection of block diagram as shown in Figure 4-23.

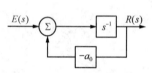
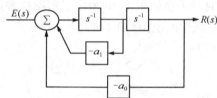

Figure 4-22　the block diagram of the first-order differential equation

Figure 4-23　the block diagram of the second-order differential equation

For general second-order differential equation
$$r''(t)+a_1 r'(t)+a_0 r(t)=b_1 e'(t)+b_0 e(t) \quad (4-82)$$
in order to represent this system by the imitation block diagram, we must introduced a complementary function $x(t)$, and let
$$e(t)=x''(t)+a_1 x'(t)+a_0 x(t) \quad (4-83)$$
substituting Eq. (4-83) into Eq. (4-82) yields
$$r''(t)+a_1 r'(t)+a_0 r(t)=b_1 (x''(t)+a_1 x'(t)+a_0 x(t))'+b_0 (x''(t)+a_1 x'(t)+a_0 x(t))$$
$$=[b_1 x'(t)+b_0 x(t)]''+a_1 [b_1 x'(t)+b_0 x(t)]'+a_0 [b_1 x'(t)+b_0 x(t)]$$
so that, we arrive at the conclusion
$$r(t)=b_1 x'(t)+b_0 x(t) \quad (4-84)$$

Combining the Eq. (4-83) and Eq. (4-84), we get the imitation block diagram as shown in Figure 4-24.

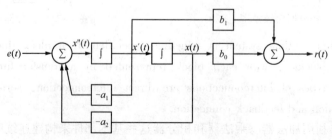

Figure 4-24　the block diagram of general second-order differential equation

For general *nth*-order linear differential equation
$$r^{(n)}(t)+a_{n-1}r^{(n-1)}(t)+\cdots+a_1r'(t)+a_0r(t)=b_m e^{(m)}(t)+b_{m-1}e^{(m-1)}(t)+\cdots+b_1e'(t)+b_0e(t)$$
(4-85)

we can use the block diagram of Figure 4-25 to imitate the *nth* order system.

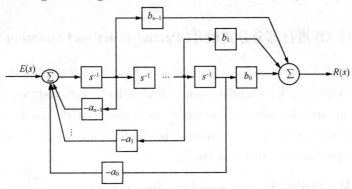

Figure 4-25 the block diagram of *nth*-order linear differential equation

Example 4-32: Consider the causal continuous-time LTI system described by the differential equation

$$\frac{d^2r(t)}{dt^2}+3\frac{dr(t)}{dt}+2r(t)=\frac{de(t)}{dt}$$

draw the imitation block diagram for this system.

Solution: According to the Eq. (4-82) and the Figure 4-24, we can draw the imitation block diagram directly as shown in Figure 4-26.

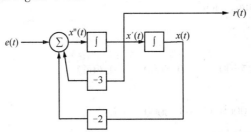

Figure 4-26 the imitation block diagram of Example 4-32

Example 4-33: Consider the causal continuous-time LTI system described by the imitation block diagram in *s*-domain, as shown in Figure 4-27. find the differential equation of this system.

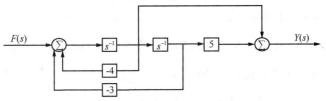

Figure 4-27 the imitation block diagram of Example 4-33

Solution: According to the Eq. (4-85) and the Figure 4-25, we can obtain the differential equation

$$\frac{d^2y(t)}{dt^2}+4\frac{dy(t)}{dt}+3y(t)=\frac{df(t)}{dt}+5f(t)$$

4.8 利用 MATLAB 进行连续时间系统的复频域分析(MATLAB Analysis of Continuous – time System in Complex Frequency Domain)

4.8.1 用 MATLAB 进行部分分式展开(Partial Fraction Expansion by MATLAB)

We can get partial fraction expansion expression of $F(s)$ by using MATLAB.

The command is [r, p, k] = residue(num, den). Here, num and den are the coefficient vectors of numerator and denominator polynomial respectively. r is the coefficient of the partial fraction. p is the poles(the roots of the denominator polynomial). k is the coefficient of polynomial. If $F(s)$ is rational proper fraction, then k is blank.

Example 4-34: Find the inverse Laplace transform of $F(s) = \dfrac{s+2}{s(s+3)(s+1)^2}$.

Solution: $F(s)$ can be written as

$$F(s) = \frac{s+2}{s(s+3)(s+1)^2} = \frac{s+2}{s^4+5s^3+7s^2+3s}$$

%Example 4-34
num = [1 2]
den = [1 5 7 3 0]
[r, p] = residue(num, den)

The running results as follows:
r =
   0.0833   -0.7500   -0.5000   0.6667
p =
   -3.0000   -1.0000   -1.0000   0

so that, $F(s)$ can be expanded as

$$F(s) = \frac{0.0833}{s+3} + \frac{-0.7500}{s+1} + \frac{-0.5000}{(s+1)^2} + \frac{0.6667}{s}$$

The inverse transform $f(t)$ is

$$f(t) = [0.08333e^{-3t} - 0.7500e^{-t} - 0.5000te^{-t} + 0.6667]u(t)$$

we obtain the same result as Example 4-20. That is,

$$f(t) = \left[\frac{1}{12}e^{-3t} - \frac{3}{4}e^{-t} - \frac{1}{2}te^{-t} + \frac{2}{3}\right]u(t)$$

Example 4-35: Find the inverse transform of the function

$$F(s) = \frac{2s^2+6s+6}{s^2+3s+2}$$

Solution:

```
%Example 4-35
num = [ 2 6 6 ]
den = [ 1 3 2 ]
k = 2
[ r, p, k ] = residue( num, den)
```

The running results as follows:

r =
    -2   2

p =
    -2   -1

k =
    2

because $F(s)$ is not the proper fraction, $F(s)$ can be expanded as

$$F(s) = \frac{2s^2+6s+6}{s^2+3s+2} = 2 + \frac{2}{s^2+3s+2} = 2 + \frac{-2}{s+2} + \frac{2}{s+1}$$

thus,

$$f(t) = \mathscr{L}^{-1}[F(s)] = 2\delta(t) - 2e^{-2t}u(t) + 2e^{-t}u(t)$$

this result by MATLAB is found to be the same as Example 4-17.

4.8.2 系统的零极点与系统特性 (Zeros and Poles of System and System Characteristic Analyzed by MATLAB)

The zeros-poles graph can be obtained from the command: pzmap(sys).

LTI system models can be obtained from the command: sys = tf(b, a). Here b and a are the coefficient vector of numerator and denominator polynomial respectively.

The impulse response $h(t)$ and the frequency response $H(j\omega)$ are obtained via the commands:
h = impulse(num, den, t);
[H, w] = freqs(num, den);

Example 4-36: Given the LTI system function $H(s) = \dfrac{s+2}{(s+5)(s^2+2s+2)}$, plot the zero-pole graph of the system and the impulse response by MATLAB, and using MATLAB to find the magnitude and phase response of the system and to judge whether the system is stable or not.

Solution:

```
%Example 4-36
num = [ 1 2 ];
den = [ 1 7 12 10 ];
sys = tf( num, den);
poles = roots( den);
```

```
subplot(2, 2, 1); pzmap(sys);
hold on
t=0 : 0.02 : 10;
h=impulse(num, den, t);
subplot(2, 2, 2); plot(t, h);
title('Impluse Respone')
hold on
w=0 : 0.01 : 1;
[H, w]=freqs(num, den);
subplot(2, 2, 3); plot(w, abs(H));
xlabel('\ \ omega')
title('Magnitude Respone')
hold on
subplot(2, 2, 4);
plot(w, angle(H));
xlabel('\ \ omega')
title('Phase Respone')
hold on
```

The running results are shown in Figure 4-28.

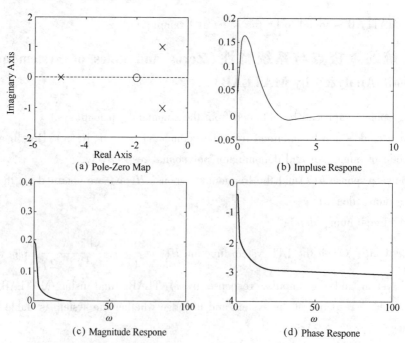

Figure 4-28 the running results of Example 4-36 by MATLAB

As can be seen from the zero-pole graph, all the poles of system are in the left half of the s-plane, therefore, the system is stable.

Example 4-37: Given the transfer function $H(s)$ of LTI circuit as follow:

$$H(s) = \frac{u_o}{u_g} = \frac{10^4(s+6000)}{s^2+875s+88\times10^6}$$

If $u_g = 12.5\cos(8000t)\,V$, find the steady state response u_o.

Solution:

%Example 4-37
w = 8000;   s = j*w;
num = [0, 1e4, 6e7];
den = [1, 875, 88e6];
H = polyval(num, s)/polyval(den, s);
mag = abs(H)
phase = angle(H)/pi * 180
t = 2 : 1e-6 : 2.002;
vg = 12.5 * cos(w*t);
vo = 12.5 * mag * cos(w*t+phase*pi/180);
plot(t, vg, t, vo); grid;
text(0.25, 0.85, 'output Voltage', 'sc');
text(0.07, 0.35, 'input Voltage', 'sc');
title('steady state filter output');
ylabel('Voltage(V)'), xlabel('time(s)');

The running resultis shown in Figure 4-29.

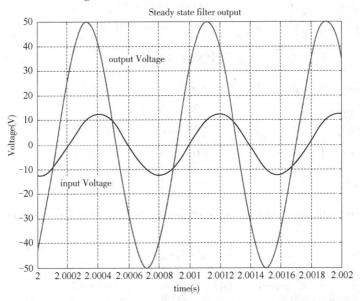

Figure 4-29   the running results of Example 4-37 by MATLAB

关键词(Key Words and Phrases)

(1) 复频域分析   Complex Frequency Domain analysis
(2) 双边拉普拉斯变换   bilateral Laplace transform

(3) 收敛域　　　　　　　　　　region of convergence
(4) 单边拉普拉斯变换　　　　　unilateral Laplace transform
(5) 拉氏变换对　　　　　　　　Laplace Transform Pairs
(6) 因果信号　　　　　　　　　causal signal
(7) 因果系统　　　　　　　　　causal system
(8) 部分分式法　　　　　　　　partial fraction expansion method
(9) 有理真分式　　　　　　　　rational proper fraction
(10) 分子多项式　　　　　　　 numerator polynomial
(11) 分母多项式　　　　　　　 denominator polynomial
(12) 极点　　　　　　　　　　 poles
(13) 零点　　　　　　　　　　 zeros
(14) 拉氏反变换　　　　　　　 inverse Laplace transform
(15) 电阻　　　　　　　　　　 resistor
(16) 电感　　　　　　　　　　 inductor
(17) 电容　　　　　　　　　　 capacitor
(18) 频率响应　　　　　　　　 frequency response
(19) 系统函数　　　　　　　　 system function
(20) 传递函数　　　　　　　　 transfer function
(21) 系统特性　　　　　　　　 system characteristic
(22) 零极点分布　　　　　　　 zeros-poles distribution
(23) 稳定系统　　　　　　　　 stable system
(24) 模拟框图　　　　　　　　 imitation block diagram

Exercises

4-1 Determine the Laplace Transform for the following signals.

(1) $(1-e^{-at})u(t)$ 　　　　　　　(2) $e^{-3t}\sin 2t\, u(t)$

(3) $t \cdot e^{-2t}u(t)$ 　　　　　　　(4) $\delta(t-t_0)+\delta'(t)$

(5) $(t-1)[u(t-1)-u(t-2)]$ 　(6) $e^{-t}[u(t)-u(t-1)]$

(7) $tu(2t-1)$ 　　　　　　　　　(8) $\dfrac{d}{dt}[e^{-t}u(t)]$

4-2 Determine the time signals corresponding to the following Laplace transform.

(1) $\dfrac{3s}{s^2+6s+8}$ 　　$\sigma>-2$ 　　(2) $\dfrac{1}{s^2-3s+2}$ 　　$\sigma>2$

(3) $\dfrac{s+3}{(s+1)^3(s+2)}$ 　　$\sigma>-1$ 　　(4) $\dfrac{2s+1}{s^2+5s+6}$ 　　$-3<\sigma<-2$

(5) $\dfrac{s+1}{(s+3)(s^2+7s+12)}$ 　　$-4<\sigma<-3$ 　　(6) $\dfrac{s+1}{(s+1)^2+4}$ 　　$\sigma<-1$

(7) $\dfrac{1-e^{-s}}{s^2+1}$ 　　$\sigma>0$ 　　(8) $\dfrac{se^{-2s}}{s^2+5s+6}$ 　　$\sigma>-2$

4-3 Find the inverse of the bilateral Laplace transform of $F_b(s)$ in its ROC, $F_b(s)$ is

given as
$$F_b(s) = \frac{6s^2+2s-2}{s(s-1)(s+2)}$$

4-4 Consider the signal $f(t) = e^{-5t} u(t-1)$, and denote its Laplace transform by $F(s)$. Evaluate $F(s)$ and specify its region of convergence.

4-5 Determine the $r(t)$ by using Laplace transform. The LTI system is given by the differential equation $\frac{d^2 r(t)}{dt^2} + 3\frac{d}{dt}r(t) + 2r(t) = \frac{de(t)}{dt} + e(t)$, the input signal $e(t) = (2e^{-t}-1)u(t)$, and the initial condition $r(0^-) = r'(0^-) = 1$.

4-6 Given the differential equation of an LTI system
$$\frac{d^2 r(t)}{dt^2} + 5\frac{d}{dt}r(t) + 6r(t) = 8\frac{de(t)}{dt} + 2e(t),$$
the input signal $e(t) = e^{-2t}u(t)$, and the initial conditions are $r(0^-) = 0$, $r'(0^-) = 1$. Determine the complete response $r(t)$ by using Laplace transform analysis method.

4-7 Determine the input signal $f(t)$ with the zero state output $y(t) = (e^{-2t} + e^{-3t})u(t)$, and the step response of this system is $g(t) = (1-e^{-2t})u(t)$.

4-8 For an LTI system, when the input signal $e(t) = e^{-2t}u(t)$, the zero state response $y_{zs}(t) = (\frac{1}{2}e^{-t} - e^{-2t} + e^{-3t})u(t)$, determine the impulse response of the system.

4-9 An LTI system described by the differential equation
$$\frac{d^2 y(t)}{dt^2} - 2\frac{dy(t)}{dt} - 8y(t) = 2\frac{df(t)}{dt} + 3f(t)$$

Find: (1) The system function $H(s)$ and plot the zeros-poles graph.

(2) The impulse response if assuming that the system is causal.

(3) The impulse response if assuming that the system is stable.

4-10 An absolutely integrable signal $f(t)$ is known to have a pole at $s = 2$. Answer the following questions:

(1) Could $f(t)$ be of finite duration?

(2) Could $f(t)$ be left sided?

(3) Could $f(t)$ be right sided?

(4) Could $f(t)$ be two sided?

4-11 The poles and zeros of an LTI continuous-time system are $p_1 = 0$, $p_2 = -1$ and $z = 1$ respectively, and the final value of the impulse response $h(\infty) = -10$, find:

(1) The system function $H(s)$.

(2) The differential equation of the system.

(3) Draw the imitation block graph of the system.

(4) The frequency response $H(j\omega)$.

4-12 Given the system functions $H(s)$ of an LTI system, please draw the zeros-poles graghs, determine the impulse responses and point out the stability of these systems.

(1) $\frac{s}{s+2}$ (2) $\frac{s+1}{s^2+2s+2}$

(3) $\dfrac{s^2+2s-3}{s^2+7s+10}$ (4) $\dfrac{s-3}{s(s+1)(s+2)}$

4-13 Consider two right-sided signals $x(t)$ and $y(t)$ related through the differential equations

$$\dfrac{dx(t)}{dt}=-2y(t)+\delta(t)$$

and

$$\dfrac{dy(t)}{dt}=2x(t)$$

determine $Y(s)$ and $X(s)$ along with their ROC.

4-14 Give the zeros-poles graph of a continuous-time system as shown in Figure 4-30 and $H(0)=\dfrac{5}{3}$. Determine the system function $H(s)$ and find the amplitude and phase response.

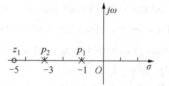

Figure 4-30   the Zeros-poles graph of Exercise 4-14

4-15 Consider the causal continuous-time system described by the following imitation block diagram as shown in Figure 4-31, find:

(1) The differential equation of the system.

(2) System function $H(s)$ and the impulse response of the system $h(t)$.

(3) Whether the system is stable and why?

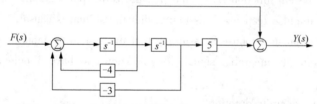

Figure 4-31   the imitation block diagram of Exercise 4-15

4-16 Given the continuous-time signal expression in s-domain $F(s)=\dfrac{s-2}{s(s+1)^3}$, find the inverse Laplace transform $f(t)$ by MATLAB.

4-17 Given the LTI system function $H(s)=\dfrac{s+2}{s^3+2s^2+2s+1}$, plot the zeros-poles graph and the step response of the system by MATLAB, and using MATLAB to find the impulse response and the amplitude and phase response.

第5章 离散时间系统的时域分析
（Analysis of the Discrete-time Systems in Time Domain）

第2~4章分别讨论了连续时间系统的时域、频域和复频域分析，本章讨论离散时间系统的时域分析。离散时间系统与连续时间系统相比有很多相似性，例如系统的线性特性、因果性、稳定性等的判别方法都是一致的；对离散时间系统的分析在很大程度上也与连续时间系统的分析相类似，例如，离散系统的全响应也可分解为零输入响应和零状态响应，利用激励信号与单位脉冲响应卷积（此处为卷积和）的方法可求零状态响应等都与求解连续系统时相一致。因此，前面章节中对连续时间系统的分析方法为离散时间系统的分析提供了便利条件。

5.1 离散时间系统的描述——差分方程（Description of the Discrete-time Systems——Difference Equation）

为了研究离散系统的性能，需要建立数学模型。连续系统可用微分方程描述，而离散系统可用差分方程描述。本节主要讨论线性离散系统差分方程的描述。

5.1.1 线性常系数差分方程（Linear Constant Coefficient Difference Equation）

一个 N 阶线性常系数差分方程的一般形式为

$$y(n) = \sum_{j=0}^{M} b_j x(n-j) - \sum_{i=1}^{N} a_i y(n-i) \tag{5-1}$$

或者

$$\sum_{i=0}^{N} a_i y(n-i) = \sum_{j=0}^{M} b_j x(n-j) \tag{5-2}$$

式中 $a_i(i=1, 2, \cdots, N)$，$b_j(j=1, 2, \cdots, M)$ 为各项系数，$a_0 = 1$。$x(n)$，$y(n)$ 分别为离散时间系统的输入和输出信号。N 为差分方程的阶数。这种形式的差分方程，各未知序列的序号自 n 以递减方式给出，称为后向差分方程（backward difference equation）。也可以从 n 以递增方式给出，即

$$\sum_{i=0}^{N} a_i y(n+i) = \sum_{j=0}^{M} b_j x(n+j) \tag{5-3}$$

称为前向差分方程（forward difference equation）。两者并没有本质上的区别，但因为前向差分方程是物理不可实现系统，所以本书讨论的是后向差分。

Difference operator of sequence is classified into the forward difference and backward difference. Because the systems described by the forward difference equations can't be realize, so we focus on the backward difference in this book.

5.1.2 微分方程与差分方程之间的关系(Relationship between Difference Equation and Differential Equation)

对于一个连续的时间系统,如果将所有时间量进行抽样,那么连续系统变成离散系统,此时微分方程也变成了差分方程。

假设微分方程为 $\frac{dy(t)}{dt}+ay(t)=bx(t)$,若将连续变量 t 等分,则有 $t=nT_s$, ($n=0$, 1, 2, …),其中 T_s 为步长。连续函数 $y(t)$ 在 $t=nT_s$ 各点的取值构成离散序列 $y(nT_s)$。在 T_s 足够小的情况下,$\frac{dy(t)}{dt}$ 可表示为

$$\frac{dy(t)}{dt} \approx \frac{y(nT_s)-y[(n-1)T_s]}{T_s} \tag{5-4}$$

代入微分方程得

$$y(nT_s)-\frac{1}{1+aT_s}y[(n-1)T_s]=\frac{bT_s}{1+aT_s}x(nT_s) \tag{5-5}$$

取 T_s 为单位时间,则

$$y(n)-\frac{1}{1+a}y(n-1)=\frac{b}{1+a}x(n) \tag{5-6}$$

这是一个一阶线性常系数差分方程。

同理,可以从高阶微分方程得到高阶差分方程。

Note: the premise condition of difference equation from the differential equation approximation is that the time interval T_s is small enough. The smaller T_s is, the better approximate degree will be.

5.1.3 离散时间系统的模拟(Imitation of the Discrete-time Systems)

差分方程和微分方程一样,既可以由实际问题的物理意义直接建立输入、输出关系,也可以通过系统的模拟框图直接建立差分方程模型。

在连续时间系统的框图描述中,有三个基本单元,分别是加法器(accumulator),放大器(multiplier)和积分器(integrator)。与之相对应,在离散时间系统的框图描述中,也有三个基本运算单元,分别为加法器、放大器和延时器(unit delayer)。加法器和放大器的描述符号与连续系统相一致,延时器在时域中用符号"D"来表示,在复频域中用"z^{-1}"表示。下面以实例说明如何在差分方程与模拟框图间进行相互转换。

For first order difference equation

$$y(n+1)+ay(n)=x(n)$$

we can rewrite as

$$y(n+1)=x(n)-ay(n)$$

Similar to the imitation of continuous-time systems, we have the imitation block diagram shown in Figure 5-1:

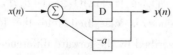

Figure 5-1 the imitation diagram of first order forward difference equation

Where, "D" represents unit time delay. The diagram consists of three components: unit delayers that store previous outputs, multipliers and accumulators.

For backward difference equation, for example, $y(n)=x(n)-ay(n-1)$, the imitation block diagram can be also plotted as shown in Figure 5-2.

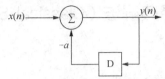

Figure 5-2 the imitation diagram of first order backward difference equation

例 5-1：对于给定的差分方程
$$y(n+2)-5y(n+1)+6y(n)=x(n+1)-2x(n)$$
将其表示为模拟框图描述方式。

解：与连续时间系统类似，对于一般二阶差分方程，我们需要引入一个辅助函数 $q(n)$，然后令输入信号
$$x(n)=q(n+2)-5q(n+1)+6q(n) \tag{5-7}$$
即
$$q(n+2)=x(n)+5q(n+1)-6q(n) \tag{5-8}$$
将式(5-7)代入差分方程得
$$y(n)=q(n+1)-2q(n) \tag{5-9}$$
根据式(5-8)和式(5-9)我们就可以画出该差分方程的模拟框图表述形式，如图 5-3 所示。

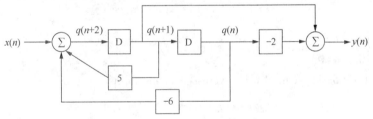

图 5-3 二阶差分方程的模拟框图

例 5-2：已知某离散时间系统的模拟框图如图 5-4 所示，写出该系统的差分方程。

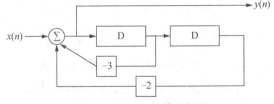

图 5-4 例 5-2 的模拟框图

解：按照例 5-1 的思路从后向前递推，可以很轻松地得到系统的差分方程为
$$y(n)+3y(n-1)+2y(n-2)=x(n)$$
也可以将其转化为前向差分形式
$$y(n+2)+3y(n+1)+2y(n)=x(n+2)$$

From above discussions, we find that the unit delayer to imitate the discrete-time system is same as the integrator to imitate thecontinuous system. So that, by using the result of the imitation for continuous-time system on former chapter, we can deduce the imitation diagram for Nth order discrete-time system.

5.2 离散时间系统的经典解法(Classical Solution of Discrete-time Systems)

如前所述,线性时不变离散系统的差分方程可以写成如下递归形式:

$$y(n) = \sum_{j=0}^{M} b_j x(n-j) - \sum_{i=1}^{N} a_i y(n-i) \tag{5-10}$$

因此,可用递归法求解 $y(n)$ 的数值。

5.2.1 递归法(Recursive Method)

由式(5-10)可知,要想求解 $y(n)$ 的值,必须要知道输入值以及 n 时刻以前的输出值。因此,当已知输入信号及 N 个初始条件 $y(-N)$, …, $y(-1)$ 时,就可以用迭代的方法求出 $y(n)$。

To illustrate the solution of such an equation, and gain some behavior and properties ofrecursive difference equations, let us examine the following simple example.

Example 5-3: Consider the difference equation

$$y(n) + 2y(n-1) = x(n) - x(n-1) \tag{5-11}$$

where, $x(n) = \begin{cases} n^2 & n \geq 0 \\ 0 & n < 0 \end{cases}$ and $y(-1) = 1$, find $y(n)$.

Solution: Eq. (5-11) can also be expressed in the form

$$y(n) = x(n) - x(n-1) - 2y(n-1)$$

let $n = 0$ then

$$y(0) = x(0) - x(-1) - 2y(-1)$$
$$= 0 - 0 - 2 = -2$$

let $n = 1, 2$ we have

$$y(1) = x(1) - x(0) - 2y(0) = 1 - 0 + 4 = 5$$
$$y(2) = x(2) - x(1) - 2y(1) = 4 - 1 - 10 = -7$$
……

显然,用迭代法求解差分方程思路清晰,便于编写计算程序,能得到方程的数值解,但不易得到解析形式的解。

5.2.2 经典解法(Classical Analysis Method)

与微分方程的经典解类似,形如上式差分方程的解由齐次解 $y_h(n)$ 和特解 $y_p(n)$ 两部分组成,即

$$y(n) = y_h(n) + y_p(n) \tag{5-12}$$

(1) 齐次解(homogeneous solution)

求齐次解就是求齐次差分方程的解。一般齐次差分方程表示为

$$\sum_{i=0}^{N} a_i y(n-i) = 0 \tag{5-13}$$

设 $\alpha_i (i=1, 2, \cdots N)$ 是特征方程 $\sum_{k=0}^{N} a_k \alpha^{N-k} = 0$ 的特征根，即满足如下方程：

$$a_0 \alpha^N + a_1 \alpha^{N-1} + \cdots + a_{N-1} \alpha^1 + a_N \alpha^0 = 0 \tag{5-14}$$

① 如果 $\alpha_i (i=1, 2, \cdots N)$ 为单实根，则齐次解为

$$y_h(n) = \sum_{i=1}^{N} A_i \alpha_i^n$$

② 如果 α_1 为 r 重根，则齐次解为

$$y(n) = (A_1 + A_2 n + A_3 n^2 + \cdots + A_r n^{r-1}) \alpha_1^n + \sum_{i=r+1}^{N} A_i \alpha_i^n$$

$$= \sum_{i=1}^{r} A_i n^{i-1} \alpha_1^n + \sum_{i=r+1}^{N} A_i \alpha_i^n$$

③ 如果特征根中有共轭复根，设 $\alpha_{1,2} = a \pm jb$，则齐次解为(仅针对共轭复根部分)

$$y_1(n) = (A_1 \cos n\varphi + A_2 \sin n\varphi) \rho^n$$

其中，$\rho = \sqrt{a^2 + b^2}$，$\varphi = \arctan \dfrac{b}{a}$。

Example 5-4: Find the homogeneous solution for the discrete-time system described by the difference equation

$$y(n) + 3y(n-1) + 2y(n-2) = x(n)$$

Solution: The homogeneous equation is

$$y(n) + 3y(n-1) + 2y(n-2) = 0$$

the characteristic equation is

$$\alpha^2 + 3\alpha + 2 = 0 \text{ or } (\alpha+2)(\alpha+1) = 0$$

so the roots are $\alpha_1 = -1$, $\alpha_2 = -2$, and homogeneous solution is

$$y_h(n) = A_1 (-1)^n + A_2 (-2)^n$$

Example 5-5: Find the homogeneous solution for the discrete-time system described by the difference equation

$$y(n) + 2y(n-1) + y(n-2) = 0$$

Solution: The characteristic equation is

$$\alpha^2 + 2\alpha + 1 = 0 \text{ or } (\alpha+1)^2 = 0$$

so the characteristic roots $\alpha_{1,2} = -1$ are repeated 2 times roots, therefore, the homogeneous solution is

$$y_h(n) = (A_1 + A_2 n)(-1)^n$$

Example 5-6: Determine the homogeneous solution for the system described by the following difference equation

$$y(n)+2y(n-1)+2y(n-2)=x(n)$$

Solution: The characteristic equation is

$$\alpha^2+2\alpha+2=0 \text{ or } (\alpha+1)^2=-1$$

and the roots are $\alpha_{1,2}=-1\pm j=-\sqrt{2}\,e^{\mp j\frac{\pi}{4}}$

the homogeneous solution is

$$y(n)=\left(A_1\cos\frac{n\pi}{4}+A_2\sin\frac{n\pi}{4}\right)(\sqrt{2})^n$$

(2) 特解(particular solution)

特解的函数形式取决于激励的函数形式。为求得特解 $y_p(n)$，需根据差分方程右端项选择合适的特解函数形式，代入方程后求出待定系数。表 5-1 中列出了几种典型激励函数所对应的特解函数式。

Table 5-1 the forms of the particular solutions

input $x(n)$			particular solution
C			B
n^m			$B_0+B_1n+B_2n^2+\cdots+B_mn^m$
a^n	$a\neq a_i$, a is not characteristic root		Ba^n
	$a=a_i$	a_i are distinct roots	$(B_0+B_1n)a^n$
		a_i is repeat r times roots	$(B_0+B_1n+\cdots+B_rn^r)a^n$
$\sin(\omega n+\varphi)$ or $\cos(\omega n+\varphi)$			$B_1\sin\omega n+B_2\cos\omega n$

Example 5-7: Find a particular solution for the first-order discrete-time system described by the difference equation

$$y(n)-\rho y(n-1)=x(n)$$

assume that the input is $x(n)=\left(\frac{1}{2}\right)^n$.

Solution: We assume a particular solution of the form $y_p(n)=B\left(\frac{1}{2}\right)^n$. Substituting $y_p(n)$ and $x(n)$ into the given difference equation, we get

$$B\left(\frac{1}{2}\right)^n-\rho B\left(\frac{1}{2}\right)^{n-1}=\left(\frac{1}{2}\right)^n$$

$$\text{or } B-\rho B\left(\frac{1}{2}\right)^{-1}=1$$

that is

$$B=\frac{1}{1-2\rho}$$

so the particular solution is

$$y_p(n)=\frac{1}{1-2\rho}\left(\frac{1}{2}\right)^n$$

If $\rho = \frac{1}{2}$, then the particular solution has the same form as the homogeneous solution and we must assume a particular solution of the form $y_p(n) = [B_0 + B_1 n]\left(\frac{1}{2}\right)^n$. Substitute this particular solution into the difference equation,

$$(B_0 + B_1 n)\left(\frac{1}{2}\right)^n - \frac{1}{2}[B_0 + B_1(n-1)]\left(\frac{1}{2}\right)^{n-1} = \left(\frac{1}{2}\right)^n$$

or
$$B_0 + B_1 n - B_0 - B_1 n + B_1 = 1$$

then
$$B_1 = 1, \quad B_0 = 0$$

so that,
$$y_p(n) = n\left(\frac{1}{2}\right)^n$$

(3) 全解 (complete solution)

将系统的齐次解和特解相加就是方程的完全解，对于完全解中的待定系数需要利用给定的边界条件来计算求得。

Example 5-8: Calculate the complete solution for the first-order recursive system described by the difference equation

$$y(n) - \frac{1}{4}y(n-1) = x(n)$$

if the input is $x(n) = \left(\frac{1}{2}\right)^n u(n)$ and the initial condition is $y(-1) = 8$.

Solution: The homogeneous solution is

$$y_h(n) = A\left(\frac{1}{4}\right)^n$$

and the particular solution is

$$y_p(n) = B\left(\frac{1}{2}\right)^n$$

substituting $y_p(n)$ into the difference equation, we get $B = 2$. thus,

$$y(n) = A\left(\frac{1}{4}\right)^n + 2\left(\frac{1}{2}\right)^n \quad \text{for } n \geq 0 \tag{5-15}$$

The coefficient A can be obtained from the initial condition. Let $n = 0$, then the difference equation becomes

$$y(0) = x(0) + \frac{1}{4}y(-1) = 1 + \frac{1}{4} \times 8 = 3$$

we substitute $y(0) = 3$ into Eq. (5-15), that is,

$$3 = 2\left(\frac{1}{2}\right)^0 + A\left(\frac{1}{4}\right)^0$$

from which we find that $A = 1$. Therefore, we can get the complete solution

$$y(n) = 2\left(\frac{1}{2}\right)^n + \left(\frac{1}{4}\right)^n, \quad \text{for } n \geq 0$$

5.3 离散时间系统的响应 (The Response of the Discrete-time System)

采样经典法分析系统响应时，存在于连续时间系统经典法相似的问题。若系统的激励项较复杂，则难以设定相应的特解形式。此外，经典法是一种纯数学方法，无法突出系统响应的物理概念。同连续时间 LTI 系统一样，离散时间 LTI 系统的完全响应也可以看作是初始状态与输入激励分别单独作用于系统产生的响应叠加。其中，由初始状态单独作用产生的输出响应称为零输入响应，记作 $y_{zi}(n)$；而由输入激励单独作用产生的输出响应称为零状态响应，记作 $y_{zs}(n)$。因此，有

$$y(n) = y_{zi}(n) + y_{zs}(n) \tag{5-16}$$

即系统的完全响应 $y(n)$ 为零输入响应 $y_{zi}(n)$ 与零状态响应 $y_{zs}(n)$ 之和。

It is informative to express the output of a system described by difference equation as the sum of two components: one associated only with the initial conditions, another due only to the input signal. We will term the former the zero input response of the system and denote it as $y_{zi}(n)$. The later is termed the zero state response of the system and is denoted as $y_{zs}(n)$. Thus, the complete response is $y(n) = y_{zi}(n) + y_{zs}(n)$.

5.3.1 零输入响应 (Zero Input Response)

零输入响应是输入激励为零时仅由初始状态所引起的输出响应。在零输入下，描述 N 阶离散时间系统的数学模型等号右端为零，为齐次方程，即

$$\sum_{i=0}^{N} a_i y(n-i) = 0$$

故零输入响应的形式与齐次差分方程齐次解的形式一致，即

$$y_{zi}(n) = \sum_{i=1}^{N} c_i \alpha_i^n \tag{5-17}$$

其中待定系数 c_i 可由系统的初始条件计算求得（这里假设 α_i 为互异的单实根）。

Example 5-9: The system is described by the difference equation

$$y(n) + 3y(n-1) + 2y(n-2) = x(n)$$

and the initial conditions are $y(-1) = 0$, $y(-2) = \dfrac{1}{2}$. Find the zero input response.

Solution: The form of the zero input response is

$$y_{zi}(n) = c_1(-1)^n + c_2(-2)^n$$

substituting the initial conditions $y(-1) = 0$, $y(-2) = \dfrac{1}{2}$ into the above equation, we have,

$$y(-1) = -c_1 - \frac{1}{2}c_2 = 0$$

$$y(-2) = c_1 + \frac{1}{4}c_2 = \frac{1}{2}$$

that is, $c_1 = 1$, $c_2 = -2$. Thus, the zero input response is

$$y_{zi}(n) = (-1)^n - 2(-2)^n, \quad n \geq 0$$

5.3.2 零状态响应(Zero State Response)

离散系统的零状态响应是系统的初始状态为零,仅由输入信号 $x(n)$ 所产生的响应,用 $y_{zs}(n)$ 表示。在连续时间系统中,通过把激励信号分解为冲激信号的线性组合,求出每一个冲激信号单独作用于系统的冲激响应,然后把这些响应叠加,即得系统对应此激励信号的零状态响应。这个叠加的过程表现为卷积积分。在离散时间系统中,可以采用相同的原理进行分析。

In the continuous-time LTI system, through the excitation signal is decomposed into a linear combination of impulse signals, we can calculate the impulse response for each impulse signal alone in the system, and then overlay these responses, which may determine zero state response to this excitation signal. The superposition process is presented as convolution integral. In the discrete-time LTI system, the same principle can be used for analysis.

离散时间系统的任意激励信号 $x(n)$ 可以表示为单位脉冲序列的线性组合,即

$$x(n) = \cdots + x(-2)\delta(n+2) + x(-1)\delta(n+1) + x(0)\delta(n) + x(1)\delta(n-1)$$
$$+ \cdots + x(m)\delta(n-m) + \cdots$$
$$= \sum_{m=-\infty}^{\infty} x(m)\delta(n-m)$$

系统在单位脉冲序列 $\delta(n)$ 作用下的零状态响应称为单位脉冲响应,用符号 $h(n)$ 表示,则系统的零状态响应可表示为

$$y_{zs}(n) = \cdots + x(-2)h(n+2) + x(-1)h(n+1) + x(0)h(n)$$
$$+ x(1)h(n-1) + \cdots + x(m)h(n-m) + \cdots$$
$$= \sum_{m=-\infty}^{\infty} x(m)h(n-m) \tag{5-18}$$

上式称为卷积和(convolution sum),用符号记为

$$y_{zs}(n) = x(n) * h(n) \tag{5-19}$$

式(5-19)表明离散时间 LTI 系统零状态响应等于激励信号和系统单位脉冲响应的卷积和。

Example 5-10: Suppose that the impulse response of the system is $h(n) = a^n u(n)$, where $0 < a < 1$, and the input signal $x(n) = u(n)$, find the zero state response $y_{zs}(n)$.

Solution: The zero state response $y_{zs}(n)$ can be determined according to Eq. (5-19):

$$y_{zs}(n) = x(n) * h(n) = h(n) * x(n)$$
$$= \sum_{m=-\infty}^{\infty} h(m)x(n-m)$$
$$= \sum_{m=-\infty}^{\infty} a^m u(m)u(n-m)$$
$$= \sum_{m=0}^{n} a^m u(n)$$
$$= \frac{1 - a^{(n+1)}}{1-a} u(n)$$

We can see that the unit impulse response $h(n)$ must be known when solving the zero state response of discrete-time system, and we need to calculate the convolution sum of two sequences. The following sections will introduce how to solve the unit impulse response and convolution sum.

5.3.3 单位脉冲响应(Unit Impulse Response)

单位脉冲序列 $\delta(n)$ 作用于离散时间系统所产生的零状态响应称为单位脉冲响应,用符号 $h(n)$ 表示,它的作用同连续时间系统的单位冲激响应 $h(t)$ 。

因为单位脉冲序列 $\delta(n)$ 只在 $n=0$ 时取值 $\delta(n)=1$,在 n 为其他值时都是零。对于因果系统,单位脉冲序列瞬时作用后,其输入变为零,此时描述系统的差分方程变为齐次方程,而单位脉冲序列对系统的瞬时作用则转化为系统的等效初始条件,这样就把问题转化为求解齐次方程,即单位脉冲响应 $h(n)$ 可写为如下形式:

$$h(n) = \sum_{i=1}^{n} k_i \alpha_i^n u(n) \qquad (5-20)$$

式中 α_i 为对应齐次差分方程的特征根,并假设其为互异根;k_i 为待定系数。

Example 5-11: Consider the difference equation of the discrete-time system

$$y(n) - 5y(n-1) + 6y(n-2) = x(n)$$

find the unit impulse response of the system.

Solution: Characteristic equation is

$$\alpha^2 - 5\alpha + 6 = 0$$

the characteristic roots are $\alpha_1 = 2$, $\alpha_2 = 3$
so that the form of $h(n)$ is

$$h(n) = [k_1(2)^n + k_2(3)^n] u(n) \qquad (5-21)$$

When the input signal is $\delta(n)$, the output signal of the system is $h(n)$, so that the difference equation can be rewritten as

$$h(n) - 5h(n-1) + 6h(n-2) = \delta(n)$$

Note: $h(n) = 0$ for $n<0$.
when $n=0$, $h(0) - 5h(-1) + 6h(-2) = \delta(0) = 1$, thus, $h(0) = 1$,
when $n=1$, we have $h(1) - 5h(0) + 6h(-1) = \delta(1) = 0$, thus, $h(1) = 5h(0) = 5$,
substituting $h(0) = 1$ and $h(1) = 5$ into Eq. (5-21), we get $k_1 + k_2 = 1$ and $2k_1 + 3k_2 = 5$,
that is, $k_1 = -2$, $k_2 = 3$
thus the unit impulse response of the discrete-time system is

$$h(n) = (3^{n+1} - 2^{n+1}) u(n)$$

对于差分方程右端包含 $x(n)$ 及其移序项的情况,例如

$$a_0 y(n) + a_1 y(n-1) + \cdots + a_N y(n-N) = b_0 x(n) + b_1 x(n-1) + \cdots + b_M x(n-M) \qquad (5-22)$$

这时,可先求出对应 $a_0 y(n) + a_1 y(n-1) + \cdots + a_N y(n-N) = b_0 x(n)$ 的单位脉冲响应 $\hat{h}(n)$,然后再用下式求出式(5-22)所描述系统的单位脉冲响应。

$$h(n) = b_0 \hat{h}(n) + b_1 \hat{h}(n-1) + \cdots + b_M \hat{h}(n-M)$$

5.4 卷积和（The Convolution Sum）

离散时间信号卷积和是计算离散时间 LTI 系统零状态响应的有力工具，同连续时间信号的卷积积分一样重要，下面将详细介绍卷积和的计算方法与性质。

The convolution sum of discrete-time signals is a powerful tool to calculate the zero state response of discrete-time LTI system. Here we will introduce the calculation and properties of the convolution sum.

5.4.1 卷积和的计算（Calculation of the Convolution Sum）

从式(5-18)可知，卷积和被定义为

$$x(n) * h(n) = \sum_{m=-\infty}^{\infty} x(m) h(n-m) \quad (5-23)$$

式中 $x(n)$ 和 $h(n)$ 为离散时间信号，m 为卷积和的自变量。根据式(5-23)我们可以方便地计算出两个序列的卷积和。

For example, consider $x(n) = u(n) - u(n-N)$ and $h(n) = a^n u(n)$, then

$$\begin{aligned}
x(n) * h(n) &= \sum_{m=-\infty}^{\infty} x(m) h(n-m) \\
&= \sum_{m=-\infty}^{\infty} [u(m) - u(m-N)] a^{n-m} u(n-m) \\
&= a^n \left\{ \sum_{m=-\infty}^{\infty} a^{-m} [u(m)u(n-m) - u(m-N)u(n-m)] \right\} \\
&= a^n \left\{ \sum_{m=0}^{n} a^{-m} u(n) - \sum_{m=N}^{n} a^{-m} u(n-N) \right\} \\
&= a^n \left[\frac{1 - a^{-(n+1)}}{1 - a^{-1}} \right] u(n) - a^n \left[\frac{a^{-N} - a^{-(n+1)}}{1 - a^{-1}} \right] u(n-N)
\end{aligned}$$

5.4.2 卷积和的性质（Properties of the Convolution Sum）

The convolution sum operation provides a number of useful properties that are given below.

(1) 交换律（commutativity）

$$x(n) * h(n) = h(n) * x(n) \quad (5-24)$$

式(5-24)说明两信号的卷积和与次序无关。

(2) 分配律（distributivity）

$$x(n) * [h_1(n) + h_2(n)] = x(n) * h_1(n) + x(n) * h_2(n) \quad (5-25)$$

(3) 结合律（associativity）

$$x(n) * [h_1(n) * h_2(n)] = [x(n) * h_1(n)] * h_2(n) \quad (5-26)$$

(4) 位移特性（shift property）

$$x(n) * \delta(n-k) = x(n-k) \quad (5-27)$$

特殊地，当 $k=0$ 时，$x(n) * \delta(n) = x(n)$。

式(5-27)表明任意信号 $x(n)$ 与位移单位脉冲序列 $\delta(n-k)$ 卷积，其结果等于信号 $x(n)$ 本身的位移。

若 $x(n)*h(n)=y(n)$，利用位移特性还可推出

$$x(n-k)*h(n)=x(n)*h(n-k)=y(n-k) \qquad (5-28)$$

位移特性对于计算卷积和是非常有用的一个性质。

Example 5-12: Consider the signals $x(n)=\delta(n)+2\delta(n-1)$ and $h(n)=2\delta(n)+4\delta(n-3)$, please calculate the convolution sum of $x(n)*h(n)$.

Solution: According to the shift property, we have

$$\begin{aligned}x(n)*h(n) &= [\delta(n)+2\delta(n-1)]*[2\delta(n)+4\delta(n-3)] \\ &= \delta(n)*2\delta(n)+4\delta(n-1)*\delta(n)+4\delta(n)*\delta(n-3)+8\delta(n-1)*\delta(n-3) \\ &= 2\delta(n)+4\delta(n-1)+4\delta(n-3)+8\delta(n-4)\end{aligned}$$

5.4.3 利用卷积和求解零状态响应(Solving the Zero State Response by Using Convolution Sum)

由本章 5.3.2 节可知，线性时不变离散系统对任意激励信号 $x(n)$ 的零状态响应等于激励 $x(n)$ 与系统的单位脉冲响应 $h(n)$ 的卷积和。因此，求出单位脉冲响应 $h(n)$，则任意输入下的系统输出就可以用卷积和求得。

Example 5-13: The system described by the first-order difference equation is

$$y(n)-\frac{1}{4}y(n-1)=x(n)$$

find the zero state response by using convolution sum, where $x(n)=\left(\frac{1}{2}\right)^n u(n)$.

Solution: Because the unit impulse response $h(n)$ is satisfied the following equation:

$$h(n)-\frac{1}{4}h(n-1)=\delta(n)$$

and the form of $h(n)$ is

$$h(n)=k_1\left(\frac{1}{4}\right)^n u(n)$$

when $n=0$, then $h(0)-\frac{1}{4}h(-1)=\delta(0)=1$, that is, $h(0)=1$

therefore $h(0)=k_1\left(\frac{1}{4}\right)^0 u(0)=1$ or $k_1=1$

thus

$$h(n)=\left(\frac{1}{4}\right)^n u(n)$$

the zero state response of the system:

$$\begin{aligned}y_{zs}(n) &= x(n)*h(n) \\ &= \sum_{m=-\infty}^{\infty}\left(\frac{1}{4}\right)^m u(m)\cdot\left(\frac{1}{2}\right)^{n-m} u(n-m) \\ &= \left(\frac{1}{2}\right)^n \sum_{m=-\infty}^{\infty}\left(\frac{1}{2}\right)^m u(m)\cdot u(n-m) \\ &= \left(\frac{1}{2}\right)^n \sum_{m=0}^{n}\left(\frac{1}{2}\right)^m u(n) = \left(\frac{1}{2}\right)^n \frac{1-\left(\frac{1}{2}\right)^{n+1}}{1-\frac{1}{2}}u(n)\end{aligned}$$

$$= \left[2\left(\frac{1}{2}\right)^n - \left(\frac{1}{2}\right)^{2n}\right]u(n)$$

$$= \left[2\left(\frac{1}{2}\right)^n - \left(\frac{1}{2}\right)^n\right]u(n)$$

Example 5-14: An LTI cascade system is shown in Figure 5-5. The input signal is the unit step sequence $u(n)$ and the unit impulse responses are respectively $h_1(n) = \delta(n) - \delta(n-4)$ and $h_2(n) = a^n u(n)$, $|a| < 1$, determine the zero state response $y(n)$.

Figure 5-5 the block diagram of Example 5-14

Solution: The output of the first system:

$$\begin{aligned}
y_1(n) &= x(n) * h_1(n) \\
&= u(n) * [\delta(n) - \delta(n-4)] \\
&= u(n) - u(n-4) \\
&= \delta(n) + \delta(n-1) + \delta(n-2) + \delta(n-3)
\end{aligned}$$

the output of the cascade system:

$$\begin{aligned}
y(n) &= y_1(n) * h_2(n) \\
&= [\delta(n) + \delta(n-1) + \delta(n-2) + \delta(n-3)] * a^n u(n) \\
&= a^n u(n) + a^{n-1} u(n-1) + a^{n-2} u(n-2) + a^{n-3} u(n-3)
\end{aligned}$$

前面已经介绍了离散时间 LTI 系统的零输入响应、单位脉冲响应以及用卷积和求解零状态响应的方法，下面举例说明系统完全响应的求解。

Example 5-15: A discrete-time system is described by the following difference equation

$$6y(n) - 5y(n-1) + y(n-2) = x(n)$$

Compute the zero input response $y_{zi}(n)$, the zero state response $y_{zs}(n)$ and the complete response $y(n)$ when $y(-1) = -11$, $y(-2) = -49$ and $x(n) = u(n)$.

Solution: ① Compute the zero input response: characteristic equation of the difference equation is

$$6\alpha^2 - 5\alpha + 1 = 0$$

the characteristic roots are $\alpha_1 = \frac{1}{2}$, $\alpha_2 = \frac{1}{3}$, then the form of the zero input response is

$$y_{zi}(n) = c_1 \left(\frac{1}{2}\right)^n + c_2 \left(\frac{1}{3}\right)^n$$

substituting $y(-1) = -11$, $y(-2) = -49$ into the above equation yields

$$y(-1) = 2c_1 + 3c_2 = -11$$
$$y(-2) = 4c_1 + 9c_2 = -49$$

that is, $c_1 = 8$ and $c_2 = -9$, thus the zero input response is

$$y_{zi}(n) = 8\left(\frac{1}{2}\right)^n - 9\left(\frac{1}{3}\right)^n, \quad n \geq 0$$

② Compute the unit impulse response: according to the definition of the unit impulse response, $h(n)$ is satisfied with the following equation

$$6h(n) - 5h(n-1) + h(n-2) = \delta(n)$$

and the form of $h(n)$ is

$$h(n) = \left[k_1\left(\frac{1}{2}\right)^n + k_2\left(\frac{1}{3}\right)^n\right]u(n) \tag{5-29}$$

when $n=0$, $6h(0)-5h(-1)+h(-2)=\delta(0)=1$, thus, $h(0)=\dfrac{1}{6}$

when $n=1$, $6h(1)-5h(0)+h(-1)=\delta(1)=0$, thus, $h(1)=\dfrac{5}{36}$

substituting $h(0)=\dfrac{1}{6}$ and $h(1)=\dfrac{5}{36}$ into Eq. (5-29), we get

$$k_1+k_2=\frac{1}{6} \text{ and } \frac{1}{2}k_1+\frac{1}{3}k_2=\frac{5}{36}$$

that is, $k_1=\dfrac{1}{2}$, $k_2=-\dfrac{1}{3}$

therefore, the unit impulse response of the system is

$$h(n) = \left[\frac{1}{2}\left(\frac{1}{2}\right)^n - \frac{1}{3}\left(\frac{1}{3}\right)^n\right]u(n)$$

③ Compute the zero state response:

$$\begin{aligned}
y_{zs}(n) &= x(n) * h(n) = u(n) * \left[\frac{1}{2}\left(\frac{1}{2}\right)^n - \frac{1}{3}\left(\frac{1}{3}\right)^n\right]u(n) \\
&= \sum_{m=-\infty}^{\infty}\left[\frac{1}{2}\left(\frac{1}{2}\right)^m - \frac{1}{3}\left(\frac{1}{3}\right)^m\right]u(m)\cdot u(n-m) \\
&= \sum_{m=0}^{n}\left[\frac{1}{2}\left(\frac{1}{2}\right)^m - \frac{1}{3}\left(\frac{1}{3}\right)^m\right]u(n) \\
&= \left[\frac{1}{2} - \frac{1}{2}\left(\frac{1}{2}\right)^n + \frac{1}{6}\left(\frac{1}{3}\right)^n\right]u(n)
\end{aligned}$$

④ The complete response:

$$y(n) = y_{zi}(n) + y_{zs}(n) = \left[\frac{1}{2} + \frac{15}{2}\left(\frac{1}{2}\right)^n - \frac{53}{6}\left(\frac{1}{3}\right)^n\right]u(n)$$

同连续时间系统一样,离散系统的完全响应(complete response)除了从系统输入的角度分解为零输入响应(zero input response)与零状态响应(zero state response)之外,还可以从其他角度分解为自由响应(自然响应)与强制响应(受迫响应)。系统的自由响应(natural response)是指完全响应 $y(n)$ 中那些与系统特征根相对应的响应,而系统强制响应(forced response)是指完全响应 $y(n)$ 中那些与外部激励相同形式的响应。除此之外,系统完全响应 $y(n)$ 还可以分解为瞬态响应与稳态响应之和。所谓瞬态响应(transient response)是指完全响应 $y(n)$ 中随时间增长而趋于零的项,而系统的稳态响应(steady response)是指完全响应 $y(n)$ 中随时间增长不趋于零的项。各种响应之间既有区别又有联系。系统的零输入响应全部属于系统的自由响应,系统的零状态响应既有系统的自由响应又含有强制响应。

For example, in Example 5-15, the natural response is $\left[\dfrac{15}{2}\left(\dfrac{1}{2}\right)^n - \dfrac{53}{6}\left(\dfrac{1}{3}\right)^n\right]u(n)$, and the forced response is $\dfrac{1}{2}u(n)$. The transient response is $\left[\dfrac{15}{2}\left(\dfrac{1}{2}\right)^n - \dfrac{53}{6}\left(\dfrac{1}{3}\right)^n\right]u(n)$, and the steady response is $\dfrac{1}{2}u(n)$.

5.5 利用 MATLAB 进行离散时间系统的时域分析（Using MATLAB to Analyse Discrete-time Systems in Time Domain）

5.5.1 单位脉冲响应的 MATLAB 实现（MATLAB Realization of Unit Impulse Response）

MATLAB 为用户提供了专门用于求离散系统单位脉冲响应,并绘制其时域波形的函数 impz()。在调用 impz() 时,与连续系统一样,我们也需要用向量来对离散系统进行表示。

设描述离散系统的差分方程为

$$\sum_{i=0}^{N} a_i y(n-i) = \sum_{j=0}^{M} b_j x(n-j)$$

则我们可以用向量 **a** 和 **b** 表示该系统,即

$$\boldsymbol{a} = [a_0, a_1, \cdots, a_{N-1}, a_N]$$
$$\boldsymbol{b} = [b_0, b_1, \cdots, b_{M-1}, b_M]$$

For example: Difference equation of a discrete-time system is:

$$y(n) - y(n-1) - 2y(n-2) = x(n)$$

then coefficient vectors **a** and **b** are written respectively:

$$\boldsymbol{a} = [1, -1, -2]; \qquad \boldsymbol{b} = [1];$$

Note: Zero must be patched those blank elements in a coefficient vector. For example, difference equation is shown as below:

$$y(n) - 8y(n-2) = x(n) - x(n-1)$$

then coefficient vectors **a** and **b** are written respectively:

$$\boldsymbol{a} = [1, 0, -8]; \qquad \boldsymbol{b} = [1, -1];$$

函数 impz() 能够绘出向量 **a** 和 **b** 定义的离散系统在指定时间范围内单位响应的时域波形,并能求出系统单位响应在指定时间范围内的数值解。impz() 函数有如下几种调用格式:

(1) impz(b, a)

This command will draw the time domain waveform of unit impulse response with the default mode.

Example 5-16: Difference equation of a discrete-time system is

$$y(n) - y(n-1) + 0.9y(n-2) = x(n)$$

Draw the waveform of unit impulse response $h(n)$ with MATLAB.

Solution: Run the following MATLAB commands:

a = [1 -1 0.9];
b = [1];
impz(b, a)

the time domain waveform of unit impulse response $h(n)$ is shown in Figure 5-6.

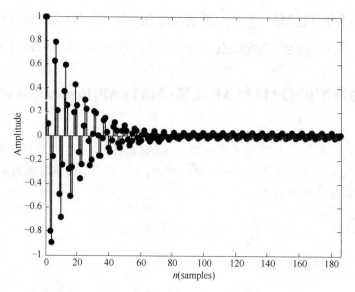

Figure 5-6 time domain waveform of unit impulse response(1)

(2) impz(b, a, n)

This command will draw the time domain waveform of unit impulse response in the range of $0 \sim n$ (n is a integer). In case of Example 5-16, if we run the following command:

impz(b, a, 40)

then the time domain waveform of unit impulse response $h(n)$ in range of $0 \sim 40$ is shown in Figure 5-7.

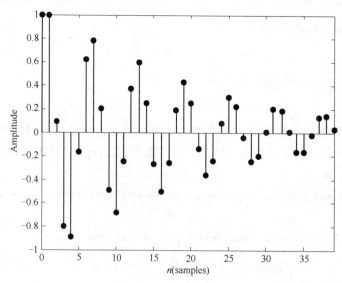

Figure 5-7 time domain waveform of unit impulse response(2)

(3) impz(b, a, n1 : n2)

This command will draw the time domain waveform of unit impulse response in the range of $n1 \sim n2$ (both $n1$ and $n2$ are integer, and $n1 < n2$). In case of Example 5-16, if we run the following command:

impz(b, a, -10: 40)

then the time domain waveform of unit impulse response $h(n)$ in range of $-10 \sim 40$ is shown in Figure 5-8.

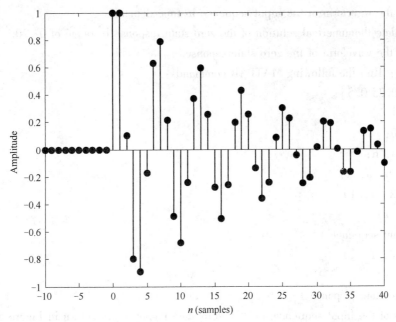

Figure 5-8　time domain waveform of unit impulse response(3)

(4) y=impz(b, a, n1:n2)

This command don't draw the time domain waveform of unit impulse response but calculate the numerical solution of unit impulse response $h(n)$ in the range of $n1 \sim n2$. In case of Example 5-16, we run the following command:

a=[1 -1 0.9];
b=[1];
y=impz(b, a, 0:10)

operation results:

y =

1.0000	1.0000	0.1000	-0.8000
-0.8900	-0.1700	0.6310	0.7840
0.2161	-0.4895	-0.6840	

5.5.2　用 MATLAB 求解离散时间系统的零状态响应(Solving the Zero State Response of the Discrete System with MATLAB)

MATLAB 为用户提供了求 LTI 离散系统响应的专用函数 filter()。该函数能求出由差分方程描述的离散系统在指定时间范围内的输入序列时所产生的零状态响应序列的数值解。

该函数的调用格式为：filter(b, a, x)

Where ***b*** and ***a*** are the coefficient vectors that describe the discrete-time system, ***x*** is a row vector including input sequence in the sample point.

Example 5-17: Difference equation of a discrete-time system is:

$$y(n)-0.25y(n-1)+0.5y(n-2)=x(n)+x(n-1)$$

and input sequence $x(n) = \left(\dfrac{1}{2}\right)^n u(n)$. Try to realize the following process with MATLAB.

① Draw the waveform of the input sequence in time domain.

② Calculate the numerical solution of the zero state response in range of 0~20.

③ Draw the waveform of the zero state response.

Solution: Run the following MATLAB commands:

```
a=[1 -0.25 0.5];
b=[1 1];
n=0:20;
x=(1/2).^n;
y=filter(b, a, x)
subplot(2, 1, 1)
stem(n, x)
title('input sequence')
subplot(2, 1, 2)
stem(n, y)
title('zero state response')
```

the waveforms of the input sequence and the zero state response are shown in Figure 5-9.

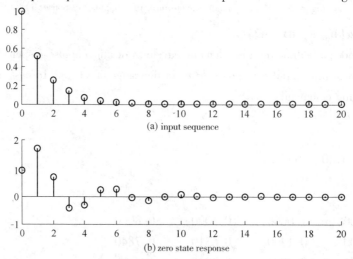

Figure 5-9   waveforms of the input sequence and the zero state response

the numerical solution of the zero state response:

y =

1.0000	1.7500	0.6875	-0.3281	-0.2383	0.1982	0.2156
-0.0218	-0.1015	-0.0086	0.0515	0.0187	-0.0204	-0.0141
0.0069	0.0088	-0.0012	-0.0047	-0.0006	0.0022	0.0008

Example 5-18: Difference equation of a discrete-time system is:

$$y(n) + y(n-1) + 0.25y(n-2) = x(n)$$

Try to draw the time domain waveform of unit step response with MATLAB.

Solution: Run the following MATLAB commands:

a=[1 1 0.25];

```
b=[1];
n=0:20;
x=ones(1,length(n));
y=filter(b,a,x);
stem(n,y)
title('unit step response')
xlabel('n');
ylabel('g(n)')
```

the time domain waveform of unit step response is shown in Figure 5-10.

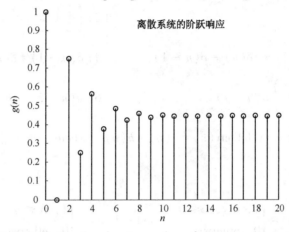

Figure 5-10　time domain waveform of unit step response

关键词(Key Words and Phrases)

(1) 离散时间系统　　　　　　　　　discrete-time system
(2) 时域　　　　　　　　　　　　　time domain
(3) 差分方程　　　　　　　　　　　difference equation
(4) 前向差分　　　　　　　　　　　forward difference
(5) 后向差分　　　　　　　　　　　backward difference
(6) 加法器　　　　　　　　　　　　accumulator
(7) 放大器　　　　　　　　　　　　multiplier
(8) 延时器　　　　　　　　　　　　unit delayer
(9) 模拟框图　　　　　　　　　　　imitation block diagram
(10) 递归法　　　　　　　　　　　 recursive method
(11) 齐次解　　　　　　　　　　　 homogeneous solution
(12) 特解　　　　　　　　　　　　 particular solution
(13) 全解　　　　　　　　　　　　 complete solution
(14) 特征方程　　　　　　　　　　 characteristic equation
(15) 特征根　　　　　　　　　　　 characteristic root
(16) 卷积和　　　　　　　　　　　 convolution sum
(17) 级联系统　　　　　　　　　　 cascade system

Exercises

5-1 Compute the complete solution of the systems described by following difference equation with input signal and initial conditions as specified.

(1) $y(n) - \frac{1}{2}y(n-1) = 2x(n)$, $y(-1) = 3$, $x(n) = \left(-\frac{1}{2}\right)^n u(n)$

(2) $y(n) - \frac{1}{9}y(n-2) = x(n-1)$, $y(-1) = 1$ $y(-2) = 0$ $x(n) = u(n)$

5-2 Calculate the convolution sum of the following sequences.

(1) $u(n) * u(n)$

(2) $\left(\frac{1}{2}\right)^n u(n) * u(n)$

(3) $\left(\frac{1}{2}\right)^n u(n-2) * [\delta(n) - \delta(n-1)]$

(4) $\delta(n-1) * \delta(n-2)$

(5) $\left(\frac{1}{2}\right)^n u(n) * \frac{1}{2}\delta(n-1)$

(6) $u(n-3) * \delta(3-n)$

5-3 Compute the convolution $y(n) = x(n) * h(n)$, where $x(n) = \left(\frac{1}{3}\right)^{-n} u(-n-1)$ and $h(n) = u(n-1)$.

5-4 Let

$$x(n) = \begin{cases} 1, & 0 \leq n \leq 9 \\ 0, & \text{otherwise} \end{cases} \quad \text{and} \quad h(n) = \begin{cases} 1, & 0 \leq n \leq N \\ 0, & \text{otherwise} \end{cases}$$

where $N \leq 9$ is an integer. Determine the value of N, given that $y(n) = x(n) * h(n)$ and $y(4) = 5$, $y(14) = 0$

5-5 For each of the following statements, determine whether it is true or false:

(1) If $x(n) = 0$ for $n < N_1$ and $h(n) = 0$ for $n < N_2$, then $x(n) * h(n) = 0$ for $n < N_1 + N_2$.

(2) If $y(n) = x(n) * h(n)$, then $y(n-1) = x(n-1) * h(n-1)$.

(3) If $y(n) = x(n) * \delta(n)$, then $y(n-2k) = x(n-k) * \delta(n-k)$.

5-6 Compute the unit impulse response $h(n)$ for each of the following discrete-time systems.

(1) $y(n) + 0.6y(n-1) = x(n)$

(2) $y(n) + 0.6y(n-1) = 2x(n-1)$

(3) $y(n) - \frac{3}{4}y(n-1) + \frac{1}{8}y(n-2) = x(n)$

(4) $y(n) - 5y(n-1) + 6y(n-2) = x(n) - 3x(n-2)$

5-7 Consider a causal system whose input and output are related by the difference equation

$$y(n) = \frac{1}{4}y(n-1) + x(n)$$

determine $y(n)$ if $x(n) = \delta(n-1)$.

5-8 The difference equation of a discrete-time system is

$$y(n) - \frac{5}{6}y(n-1) + \frac{1}{6}y(n-2) = x(n)$$

and $y(-1) = 0$, $y(-2) = 1$, $x(n) = u(n)$, please compute:

(1) The unit impulse response $h(n)$.

(2) The zero input response $y_{zi}(n)$, the zero state response $y_{zs}(n)$ and the complete response $y(n)$.

5-9 A discrete-time system has the following difference equation:
$$y(n)-3y(n-1)+2y(n-2)=x(n)-2x(n-1)$$
and the initial conditions are $y(-2)=2$, $y(-1)=1$, the input is $x(n)=u(n)$.

(1) Find the $y_{zi}(n)$, $y_{zs}(n)$ and $y(n)$.

(2) Plot the imitation graph of the system.

5-10 Given the difference equation of an LTI discrete-time system as following:
$$y(n)-4y(n-1)+4y(n-2)=4x(n)$$
and $y(-1)=0$, $y(-2)=2$, $x(n)=(-3)^n u(n)$, please determine:

(1) The unit impulse response $h(n)$.

(2) The zero input response $y_{zi}(n)$, the zero state response $y_{zs}(n)$ and the complete response $y(n)$.

(3) If $x(n)=5(-3)^{n-1}u(n-1)$, please recalculate (1) and (2).

5-11 A discrete-time system is shown in Figure 5-11. Let initial condition is zero. Write out the difference equation and find the unit impulse response.

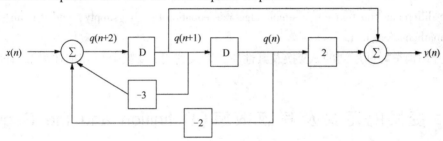

Figure 5-11 the imitation block of a discrete-time system

5-12 An LTI cascade system is shown in Figure 5-12. The input signal is the unit step sequence $u(n)$ and the unit impulse responses are respectively $h_1(n)=\delta(n)-\delta(n-3)$ and $h_2(n)=(0.8)^n u(n)$, determine the zero state response $y(n)$ respectively according to the following formula:

(1) $y(n)=[x(n)*h_1(n)]*h_2(n)$

(2) $y(n)=x(n)*[h_1(n)*h_2(n)]$

Figure 5-12 LTI cascade system

5-13 Given the difference equation of discrete-time system as following, try to calculate the numerical solution of the unit impulse response $h(n)$ and unit step response $g(n)$, and draw the time domain waveform of $h(n)$ and $g(n)$ with MATLAB.

(1) $y(n)+2y(n-1)=x(n-1)$

(2) $y(n)+y(n-1)+0.5y(n-2)=x(n)+x(n-1)$

第6章 离散时间信号与系统的 z 域分析
(Analysis of Discrete-time Signals and Systems in z-Domain)

在连续时间信号与系统中,其变换域分析方法是傅里叶变换与拉普拉斯变换。而在离散时间信号与系统中,变换域分析法中最重要的一种就是 z 变换法,它在离散时间信号与系统中的作用就如同拉普拉斯在连续时间信号与系统中的作用一样,是把描述离散系统的差分方程转化为简单的代数方程,使其求解大为简化。因而对求解离散系统而言,z 变换是一个极其重要的数学工具。z 变换的概念可以从理想抽样信号的拉普拉斯变换引出,也可以在离散域直接给出。

In the continuous-time signals and systems, its analysis method in the transform domain is the Fourier Transform and the Laplace transform, while in the discrete-time signals and systems, one of the most important transform domain analysis method is the z-transform. The z-transform can convert the difference equation into a simple algebraic equation to solve simply, and it is an extremely important mathematical tool.

本章介绍离散时间序列的 z 变换及其性质,以及利用 z 变换分析和描述离散时间系统的方法。

6.1 z 变换的定义及其收敛域(Definition and the Region of Convergence of z-transform)

6.1.1 z 变换的定义(Definition of the z-transform)

序列 $x(n)$ 的 z 变换定义为

$$X(z) = \mathscr{Z}[x(n)] = \sum_{n=-\infty}^{\infty} x(n) z^{-n} \qquad (6-1)$$

其中,z 是复变量(complex variable),$X(z)$ 是 $x(n)$ 的双边 z 变换(bilateral z-transform)。相应地,$x(n)$ 的单边 z 变换(unilateral z-transform)定义为

$$X(z) = \mathscr{Z}[x(n)] = \sum_{n=0}^{\infty} x(n) z^{-n} \qquad (6-2)$$

显然,若 $x(n)$ 是因果序列(causal sequence),则双边 z 变换与单边 z 变换是等同的。

在拉普拉斯变换分析中着重讨论单边拉普拉斯变换,这是由于在连续时间系统中,非因果信号的应用较少。但对于离散时间系统,非因果序列也有一定的应用范围,因此,本章将着重单边 z 变换适当兼顾双边 z 变换分析。

由 z 变换的定义可以看到,只有级数收敛时 z 变换才有意义。对于任意给定的有界序列 $x(n)$,使 z 变换的定义式级数收敛的所有 z 值的集合称为 $X(z)$ 的**收敛域**。与拉普拉斯变换的情况类似,对于单边 z 变换,序列与变换式唯一对应,同时也有唯一的收敛域。而在双边

z 变换时, 不同的序列在不同的收敛域条件下可能映射为同一个变换式, 我们将在后面举例说明。

From Eq. (6-1) and Eq. (6-2), we see that, for convergence of the z-transform, we require that the series of $x(n)z^{-n}$ converge. In general, the z-transform of a sequence has associated with it a range of z for which $X(z)$ converges. This range of values is referred to as the **Region of Convergence (ROC)**.

若已知 $X(z)$ 及其收敛域, 可以根据下面的公式求 $x(n)$。

$$x(n) = \mathcal{Z}^{-1}[X(z)] = \frac{1}{2\pi j}\oint_C X(z)z^{n-1}\mathrm{d}z \qquad (6-3)$$

其中 C 是包围 $X(z)z^{n-1}$ 所有极点的逆时针闭合积分路线。式(6-3)称为序列的 z 反变换, 有关 z 反变换的求法将在本章 6.4 节中进行讨论。

6.1.2 z 变换的收敛域(The Region of Convergence of z-transform)

为了讨论 z 变换的收敛域, 下面我们举例说明。

Example 6-1: Consider the signal $x(n) = a^n u(n)$. Calculate the z-transform of $x(n)$.
Solution: From Eq. (6-1).

$$X(z) = \sum_{n=-\infty}^{\infty} a^n u(n) z^{-n} = \sum_{n=0}^{\infty} (az^{-1})^n$$

For convergence of $X(z)$, we require that $\sum_{n=0}^{\infty} |az^{-1}| < \infty$. Therefore, the region of convergence is the range of z for which $|az^{-1}| < 1$, that is, $|z| > |a|$. Then,

$$X(z) = \sum_{n=0}^{\infty} (az^{-1})^n = \frac{1}{1 - az^{-1}} = \frac{z}{z - a}, \quad |z| > |a| \qquad (6-4)$$

Thus, the z-transform for this signal is well-defined for any value of a, with an ROC determined by the magnitude of a according to Eq. (6-4). For example, for $a = 1$, $x(n)$ is the unit step sequence with z-transform

$$X(z) = \frac{z}{z - 1}, \quad |z| > 1$$

我们看到式(6-4)是一个有理函数, 因此, 与拉普拉斯变换一样, z 变换也可由该有理函数的零极点表征。此例中有一个零点 $z = 0$, 一个极点 $z = a$。例 6-1 的零极点图及其收敛域如图 6-1 所示, 其中 $0 < a < 1$。对于 $|a| > 1$, 收敛域将不包含单位圆。通常情况下, 对于一个右边序列(right-sided sequence), 其 z 变换的收敛域是一个以原点为圆心, 以 a 为半径的圆外区域($z = \infty$ 处可能例外)。

Figure 6-1 pole-zero graph and region of convergence for Example 6-1 for $0 < a < 1$

We see that the z-transform in Eq. (6-4) is a rational function. Consequently, just as with rational Laplace transform, the z-transform can be characterized by its zeros (the roots of the numerator polynomial) and its poles (the roots of the denominator polynomial). For this example, there is one zero, at $z = 0$, and one pole, at $z = a$. The pole-zero graph and region of convergence for Example 6-1 are shown in Figure 6-1 for a value of a between 0 and 1. For $|a| > 1$, the ROC does not include the unit circle.

Example 6-2: Now let $x(n) = -a^n u(-n-1)$, find its z-transform.

Solution:
$$X(z) = -\sum_{n=-\infty}^{\infty} a^n u(-n-1) z^{-n} = -\sum_{n=-\infty}^{-1} a^n z^{-n}$$
$$= -\sum_{n=1}^{\infty} a^{-n} z^n = 1 - \sum_{n=0}^{\infty} (a^{-1} z)^n \qquad (6-5)$$

If $|a^{-1} z| < 1$, or equivalently, $|z| < |a|$, the summation in Eq. (6-5) converges and
$$X(z) = 1 - \frac{1}{1-a^{-1}z} = \frac{-a^{-1}z}{1-a^{-1}z} = \frac{1}{1-az^{-1}} = \frac{z}{z-a}, \quad |z| < |a| \qquad (6-6)$$

The pole-zero graph and region of convergence for this example are shown in Figure 6-2 for a value of a between 0 and 1.

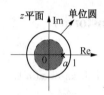

Figure 6-2 pole-zero graph and region of convergence for Example 6-2 for 0<a<1

通常情况下，对于一个左边序列(left-sided sequence)，其 z 变换的收敛域是一个以原点为圆心，以 a 为半径的圆内区域($z=0$ 处可能例外)。

比较上面两个例子，我们发现 $X(z)$ 的表达式及零极点图都相同，不同之处只在于两者的收敛域。因此，在描述 z 变换时，一定要同时给出 $X(z)$ 的表达式及其收敛域。

Comparing above two examples, we see that the algebraic expression for $X(z)$ and the corresponding pole-zero graph are identical, and the z-transform differ only in their ROC. Thus, as with the Laplace transform, specification of the z-transform requires both the algebraic expression and the region of convergence. Also, in both examples, the sequences were exponentials and the resulting z-transforms were rational. In fact, as further suggested by the following example, $X(z)$ will be rational whenever $x(n)$ is a linear combination of real or complex exponentials.

Example 6-3: Let us consider a signal that is the sum of two real exponentials:
$$x(n) = \left(\frac{1}{2}\right)^n u(n) - 2^n u(-n-1)$$

Find the z-transform of $x(n)$.

Solution:
$$X(z) = \sum_{n=0}^{\infty} \left(\frac{1}{2}\right)^n z^{-n} - \sum_{n=-\infty}^{-1} 2^n z^{-n}$$
$$= \frac{1}{1-\frac{1}{2}z^{-1}} + \frac{1}{1-2z^{-1}} \qquad (6-7)$$

For convergence of $X(z)$, both sum in Eq. (6-7) must converges, which requires that both $\left|\frac{1}{2}z^{-1}\right| < 1$ and $|2^{-1}z| < 1$, or equivalently, $|z| > \frac{1}{2}$ and $|z| < 2$. Thus, the region of convergence is $\frac{1}{2} < |z| < 2$. The pole-zero graph and ROC for the z-transform of the combined signal are shown in Figure 6-3.

通常情况下，对于一个双边序列(two-sides sequence)，其 z 变换的收敛域是某个圆环区域。

Figure 6-3 pole-zero graph and ROC for the z-transform of Example 6-3

6.2 常用序列的 z 变换（The z-transform of Basic Sequence）

6.2.1 单位脉冲序列（Unit Impulse Sequence）

$$\mathscr{Z}[\delta(n)] = \sum_{n=-\infty}^{\infty} \delta(n) z^{-n} = 1 \tag{6-8}$$

收敛域为整个 z 平面。

6.2.2 单边指数序列（Unilateral Exponential Sequence）

Given a real or complex number a, then the z-transform of $a^n u(n)$ is

$$\mathscr{Z}[a^n u(n)] = \sum_{n=0}^{\infty} a^n z^{-n} = \sum_{n=0}^{\infty} \left(\frac{a}{z}\right)^n = \frac{1}{1-\frac{a}{z}}$$

$$= \frac{z}{z-a}, \qquad |z| > |a| \tag{6-9}$$

6.2.3 单位阶跃序列（Unit Step Sequence）

From Eq. (6-9), let $a=1$, we have

$$\mathscr{Z}[u(n)] = \frac{z}{z-1}, \qquad |z| > 1 \tag{6-10}$$

6.2.4 矩形序列（Rectangular Sequence）

Let $G_N(n) = u(n) - u(n-N)$, then its z-transform is

$$\mathscr{Z}[G_N(n)] = \sum_{n=0}^{N-1} z^{-n} = \frac{1-z^{-N}}{1-z^{-1}}, \qquad |z| > 0 \tag{6-11}$$

6.2.5 单位斜变序列（Unit Ramp Sequence）

Let $x(n) = nu(n)$, then its z-transform is given by

$$X(z) = \mathscr{Z}[x(n)] = \sum_{n=0}^{\infty} n z^{-n}$$

because

$$\sum_{n=0}^{\infty} z^{-n} = \frac{1}{1-z^{-1}}, \qquad |z| > 1$$

by differentiating, we have

$$\sum_{n=0}^{\infty} -n z^{-n-1} = \frac{-z^{-2}}{(1-z^{-1})^2}$$

that is,

$$\sum_{n=0}^{\infty} n z^{-n} = \frac{z^{-1}}{(1-z^{-1})^2} = \frac{z}{(z-1)^2}$$

thus the z-transform

$$X(z) = \mathscr{Z}[nu(n)] = \frac{z}{(z-1)^2}, \qquad |z| > 1 \tag{6-12}$$

6.2.6 单边余弦与正弦序列(Unilateral Cosine and Sine Sequences)

Let $a = e^{j\omega_0}$ and $a = e^{-j\omega_0}$, then by using $\mathscr{Z}[a^n u(n)] = \dfrac{z}{z-a}$ $|z| > |a|$, we have

$$\mathscr{Z}[e^{j\omega_0 n} u(n)] = \frac{z}{z - e^{j\omega_0}}, \quad |z| > |e^{j\omega_0}| = 1$$

$$\mathscr{Z}[e^{-j\omega_0 n} u(n)] = \frac{z}{z - e^{-j\omega_0}}, \quad |z| > |e^{j\omega_0}| = 1$$

Using the above results and the Euler formula, we can get the z-transform of cosine and sine sequences,

$$\mathscr{Z}[\cos n\omega_0 u(n)] = \mathscr{Z}\left(\frac{1}{2}e^{jn\omega_0} + \frac{1}{2}e^{-jn\omega_0}\right)$$

$$= \frac{1}{2}\frac{1}{z - e^{j\omega_0}} + \frac{1}{2}\frac{1}{z - e^{-j\omega_0}}$$

$$= \frac{z(z - \cos\omega_0)}{z^2 - 2z\cos\omega_0 + 1}, \quad |z| > 1 \qquad (6-13)$$

$$\mathscr{Z}[\sin n\omega_0 u(n)] = \mathscr{Z}\left(\frac{e^{jn\omega_0} - e^{-jn\omega_0}}{2j}\right)$$

$$= \frac{1}{2j}\left(\frac{z}{z - e^{j\omega_0}} - \frac{z}{z - e^{-j\omega_0}}\right)$$

$$= \frac{z\sin\omega_0}{z^2 - 2z\cos\omega_0 + 1}, \quad |z| > 1 \qquad (6-14)$$

为了计算方便,现将常用序列的 z 变换列于表 6-1 中,以便查阅。

表 6-1　常用序列的 z 变换及其收敛域

序　列	z 变换	收　敛　域
$\delta(n)$	1	整个 z 平面
$u(n)$	$\dfrac{z}{z-1} = \dfrac{1}{1-z^{-1}}$	$|z| > 1$
$a^n u(n)$	$\dfrac{z}{z-a} = \dfrac{1}{1-az^{-1}}$	$|z| > |a|$
$u(n) - u(n-N)$	$\dfrac{1-z^{-N}}{1-z^{-1}}$	$|z| > 0$
$nu(n)$	$\dfrac{z}{(z-1)^2} = \dfrac{z^{-1}}{(1-z^{-1})^2}$	$|z| > 1$
$n^2 u(n)$	$\dfrac{z(z+1)}{(z-1)^3} = \dfrac{z^{-1}(z^{-1}+1)}{(1-z^{-1})^3}$	$|z| > 1$
$na^n u(n)$	$\dfrac{z}{(z-a)^2} = \dfrac{z^{-1}}{(1-az^{-1})^2}$	$|z| > |a|$
$\cos n\omega_0 u(n)$	$\dfrac{z(z-\cos w_0)}{z^2 - 2z\cos w_0 + 1}$	$|z| > 1$
$\sin n\omega_0 u(n)$	$\dfrac{z\sin w_0}{z^2 - 2z\cos w_0 + 1}$	$|z| > 1$

6.3 z 变换的性质（Properties of the z-transform）

z 变换有很多有用的性质，一方面可以利用这些性质求解序列的 z 变换及其反变换，另一方面也可以利用这些性质对 LTI 离散时间系统进行变换域分析。

The z-transform possesses a number of properties that are useful in deriving transform pairs and in the application of the transform to the study of LTI discrete time systems. In this section the properties of the z-transform are stated and proved.

6.3.1 线性（Linearity）

If $x_1(n) \leftrightarrow X_1(z)$, $ROC = R_1$; $x_2(n) \leftrightarrow X_2(z)$, $ROC = R_2$, then the linearity property is represented by

$$ax_1(n) + bx_2(n) \leftrightarrow aX_1(z) + bX_2(z), \qquad ROC = R_1 \cap R_2. \tag{6-15}$$

The proof of Eq. (6-15) follows directly from the definition of the z-transform. The details are omitted.

Example 6-4: Determine the z-transform of the sequence $a^n u(n) - a^n u(n-1)$.

Solution: Because $a^n u(n) \leftrightarrow \dfrac{z}{z-a}$, $|z| > |a|$ and

$$\mathscr{Z}[a^n u(n-1)] = \sum_{n=1}^{\infty} a^n z^{-n} = \frac{az^{-1}}{1-az^{-1}} = \frac{a}{z-a}, \quad |z| > |a|$$

thus,
$$\mathscr{Z}[a^n u(n) - a^n u(n-1)] = \frac{z}{z-a} - \frac{a}{z-a} = 1$$

the ROC is the whole z-plane.

如果在线性组合中某些零点与极点相抵消，则收敛域有可能扩大。

6.3.2 位移性质（Time Shifting）

序列的位移性表示序列位移后的 z 变换与原序列 z 变换的关系。在实际中可能遇到序列的左移（超前）或右移（延迟）两种不同情况，所取的变换形式可能有单边 z 变换与双边 z 变换，它们的位移性基本相同，但又各具不同的特点。下面分几种情况进行讨论。

（1）双边 z 变换（bilateral z-transform）

若序列的双边 z 变换为 $\mathscr{Z}[x(n)] = X(z)$，则序列右移后的双边 z 变换为

$$\mathscr{Z}[x(n-m)] = z^{-m} X(z) \tag{6-16}$$

Proof: According to the definition of the z-transform, we have

$$\mathscr{Z}[x(n-m)] = \sum_{n=-\infty}^{\infty} x(n-m) z^{-n} = z^{-m} \sum_{k=-\infty}^{\infty} x(k) z^{-k} = z^{-m} X(z) \tag{6-17}$$

similarly, the bilateral z-transform of left shift sequence is

$$\mathscr{Z}[x(n+m)] = z^m X(z) \tag{6-18}$$

where, m is an arbitrary positive integer.

序列位移只会使 z 变换在 $z=0$ 或 $z=\infty$ 处的零极点情况发生变化。如果 $x(n)$ 是双边序列，$X(z)$ 的收敛域为环形区域，在这种情况下序列位移并不会使 z 变换收敛域发生变化。

(2) 单边 z 变换(unilateral z-transform)

若 $x(n)$ 是双边序列，其单边 z 变换为 $\mathscr{Z}[x(n)u(n)] = X(z)$，则序列左移后的单边 z 变换为

$$\mathscr{Z}[x(n+m)u(n)] = z^m \left[X(z) - \sum_{k=0}^{m-1} x(k) z^{-k} \right] \quad (6-19)$$

Proof: According to the definition of the z-transform, we have

$$\mathscr{Z}[x(n+m)u(n)] = \sum_{n=0}^{\infty} x(n+m) z^{-n}$$

$$= z^m \sum_{n=0}^{\infty} x(n+m) z^{-(n+m)}$$

$$= z^m \left[\sum_{k=m}^{\infty} x(k) z^{-k} \right]$$

$$= z^m \left[\sum_{k=0}^{\infty} x(k) z^{-k} - \sum_{k=0}^{m-1} x(k) z^{-k} \right]$$

$$= z^m \left[X(z) - \sum_{k=0}^{m-1} x(k) z^{-k} \right]$$

similarly,

$$\mathscr{Z}[x(n-m)u(n)] = z^{-m} \left[X(z) + \sum_{k=-m}^{-1} x(k) z^{-k} \right] \quad (6-20)$$

Proof:

$$\mathscr{Z}[x(n-m)u(n)] = \sum_{n=0}^{\infty} x(n-m) z^{-n}$$

$$= z^{-m} \sum_{n=0}^{\infty} x(n-m) z^{-(n-m)}$$

$$= z^{-m} \left[\sum_{k=-m}^{\infty} x(k) z^{-k} \right]$$

$$= z^{-m} \left[\sum_{k=0}^{\infty} x(k) z^{-k} + \sum_{k=-m}^{-1} x(k) z^{-k} \right]$$

$$= z^{-m} \left[X(z) + \sum_{k=-m}^{-1} x(k) z^{-k} \right]$$

If $x(n)$ is a **causal sequence**, then

$$\mathscr{Z}[x(n-m)u(n)] = z^{-m} X(z)$$

and

$$\mathscr{Z}[x(n+m)u(n)] = z^m \left[X(z) - \sum_{k=0}^{m-1} x(k) z^{-k} \right]$$

In general, the ROC of the unilateral z-transform for shifted signal is unchanged, **except for** the possible addition or deletion of the origin or infinity.

Example 6-5: Let $x(n) = u(n) - u(n-3)$, find the z-transform of $x(n)$.

Solution: Because

$$\mathscr{Z}[u(n)] = \frac{z}{z-1}, \quad |z| > 1$$

and

$$\mathscr{Z}[u(n-3)] = z^{-3} \frac{z}{z-1} = \frac{z^{-2}}{z-1}, \quad |z| > 1$$

therefore,
$$\mathcal{Z}[x(n)] = \frac{z}{z-1} - \frac{z^{-2}}{z-1} = \frac{z^2+z+1}{z^2}$$

because $x(n)$ is a rectangular sequence, so the ROC is $|z|>0$.

6.3.3 尺度变换(Scaling in the z-Domain)

If $\mathcal{Z}[x(n)] = X(z)$, $ROC = R$, then
$$\mathcal{Z}[a^n x(n)] = X\left(\frac{z}{a}\right), \quad ROC = |a|R \tag{6-21}$$

Proof:
$$\mathcal{Z}[a^n x(n)] = \sum_{n=0}^{\infty} a^n x(n) z^{-n} = \sum_{n=0}^{\infty} x(n) \left(\frac{z}{a}\right)^{-n} = X\left(\frac{z}{a}\right)$$

especially, let $a = e^{j\omega_0}$ and $a = e^{-j\omega_0}$ then
$$\mathcal{Z}[e^{j n\omega_0} x(n)] = X(e^{-j\omega_0} z) \tag{6-22}$$
and
$$\mathcal{Z}[e^{-j n\omega_0} x(n)] = X(e^{j\omega_0} z) \tag{6-23}$$

we can use the Euler formula yields:
$$(\cos\omega_0 n) x(n) = \frac{1}{2}[e^{j\omega_0 n} x(n) + e^{-j\omega_0 n} x(n)] \leftrightarrow \frac{1}{2}[X(e^{-j\omega_0} z) + X(e^{j\omega_0} z)] \tag{6-24}$$

$$(\sin\omega_0 n) x(n) = \frac{1}{2j}[e^{j\omega_0 n} x(n) - e^{-j\omega_0 n} x(n)] \leftrightarrow \frac{1}{2j}[X(e^{-j\omega_0} z) - X(e^{j\omega_0} z)] \tag{6-25}$$

6.3.4 z 域微分性质(Differentiation in the z-Domain)

If $\mathcal{Z}[x(n)] = X(z)$, $ROC = R$, then
$$\mathcal{Z}[n x(n)] = -z \frac{dX(z)}{dz}, \quad ROC = R \tag{6-26}$$

Proof: Firstly, recall the definition of the z-transform:
$$X(z) = \sum_{n=-\infty}^{\infty} x(n) z^{-n} \tag{6-27}$$

by differentiating,
$$\frac{d}{dz} X(z) = \sum_{n=-\infty}^{\infty} (-n) x(n) z^{-n-1} = -z^{-1} \sum_{n=-\infty}^{\infty} n x(n) z^{-n}, \quad \text{with } ROC = R$$

thus
$$\mathcal{Z}[n x(n)] = \sum_{n=-\infty}^{\infty} n x(n) z^{-n} = -z \frac{d}{dz} X(z), \quad \text{with } ROC = R$$

similarly,
$$\mathcal{Z}[n^2 x(n)] = -z \frac{d}{dz}\left[-z \frac{d}{dz} X(z)\right] \tag{6-28}$$

Example 6-6: Let $x(n) = a^n u(n)$, where a is any nonzero real or complex.
Solution: From Table 6-1,
$$a^n u(n) \leftrightarrow \frac{z}{z-a}, \quad |z| > |a|$$

then
$$n a^n u(n) \leftrightarrow -z \frac{d}{dz}\left(\frac{z}{z-a}\right) = \frac{az}{(z-a)^2}, \quad |z| > |a| \tag{6-29}$$

note that when $a = 1$, Eq. (6-29) becomes

$$nu(n) \leftrightarrow \frac{z}{(z-1)^2} \tag{6-30}$$

similarly,
$$\mathscr{Z}[n^2 a^n u(n)] = \mathscr{Z}[n \cdot n a^n u(n)] = -z \frac{\mathrm{d}}{\mathrm{d}z}\left[\frac{az}{(z-a)^2}\right] = \frac{az(z+a)}{(z-a)^3} \tag{6-31}$$

letting $a=1$ in Eq. (6-31) results in the transform pair

$$n^2 u(n) \leftrightarrow \frac{z(z+1)}{(z-1)^3} \tag{6-32}$$

6.3.5 反褶序列(Time Reversal)

If $\mathscr{Z}[x(n)] = X(z)$, $ROC = R$, then

$$\mathscr{Z}[x(-n)] = X\left(\frac{1}{z}\right), \qquad ROC = \frac{1}{R} \tag{6-33}$$

as we can see, the ROC of $x(-n)$ is the reciprocal of that of $x(n)$.

Proof:
$$\mathscr{Z}[x(-n)] = \sum_{n=-\infty}^{\infty} x(-n) z^{-n} = \sum_{n=-\infty}^{\infty} x(n) z^{n}$$
$$= \sum_{n=-\infty}^{\infty} x(n) (z^{-1})^{-n}$$
$$= X\left(\frac{1}{z}\right), \ ROC = \frac{1}{R}$$

6.3.6 求和定理(Summation Theorem)

对于因果序列 $x(n)$，若 $\mathscr{Z}[x(n)] = X(z)$，$|z| > R_x$，R_x 为最小收敛半径，则

$$\mathscr{Z}\left[\sum_{i=0}^{n} x(i)\right] = \frac{1}{1-z^{-1}} X(z), \quad |z| > \max(R_x, 1) \tag{6-34}$$

Proof: Let $g(n) = \sum_{i=0}^{n} x(i) \leftrightarrow G(z)$

because
$$\sum_{i=0}^{n} x(i) - \sum_{i=0}^{n-1} x(i) = g(n) - g(n-1) = x(n)$$

taking the z-transform of the above equation on both sides, we get
$$G(z) - z^{-1} G(z) = X(z)$$

that is,
$$G(z) = \mathscr{Z}\left[\sum_{i=0}^{n} x(i)\right] = \frac{1}{1-z^{-1}} X(z), \qquad |z| > \max(R_x, 1)$$

Example 6-7: Find the z-transform of the sequence $\sum_{i=0}^{n} u(i-1)$.

Solution: Because $u(n-1) \leftrightarrow z^{-1} \frac{z}{z-1}$, $|z| > 1$

using the summation theorem, we can find easily the z-transform:

$$\mathscr{Z}\left[\sum_{i=0}^{n} u(i-1)\right] = \frac{1}{1-z^{-1}} \cdot z^{-1} \frac{z}{z-1} = \frac{z}{(z-1)^2}, \quad |z| > 1$$

6.3.7 初值定理(Initial-Value Theorem)

For a causal sequence $x(n)$, we have

$$x(0) = \lim_{z \to \infty} X(z) \qquad (6-35)$$

Proof: This property follows by considering the limit of each term individually in the expression for the z-transform, that is,

$$X(z) = \sum_{n=0}^{\infty} x(n) z^{-n} = x(0) + x(1) z^{-1} + x(2) z^{-2} + \cdots$$

as $z \to \infty$, $z^{-n} \to 0$, (for $n > 0$), whereas $z^{-n} = 1$ (for $n = 0$). Thus, Eq. (6-34) follows.

由初值定理可以看出，对一个因果序列来说，如果 $x(0)$ 是有限值的话，那么 $\lim_{z \to \infty} X(z)$ 就是有限值。如果将 $X(z)$ 表示成 z 的两个多项式之比的话，分子多项式的阶次一定小于分母多项式的阶次，或者说零点的个数不能多于极点的个数。

As one consequence of the initial-value theorem, for a causal sequence, if $x(0)$ is finite, then $\lim_{z \to \infty} X(z)$ is finite. Consequently, with $X(z)$ expressed as a ratio of polynomials in z, the order of the numerator polynomial cannot be greater than the order of the denominator polynomial; or, equivalently, the number of finite zeros of $X(z)$ cannot be greater than the number of finite poles.

6.3.8 终值定理(Final-Value Theorem)

对于因果序列 $x(n)$，若 $\mathscr{Z}[x(n)] = X(z)$ 的极点在单位圆内，且只允许单位圆上 $z=1$ 处有一极点，则有

$$\lim_{n \to \infty} x(n) = x(\infty) = \lim_{z \to 1}(z-1) X(z) \qquad (6-36)$$

Proof: If $x(n)$ is causal sequence, that is, $x(n) = 0$, $n < 0$, then

$$X(z) = \mathscr{Z}[x(n)] = \sum_{n=0}^{\infty} x(n) z^{-n}$$

and

$$\mathscr{Z}[x(n+1) - x(n)] = zX(z) - x(0) - X(z)$$
$$= (z-1) X(z) - x(0)$$

taking the limit as $z \to 1$ of both sides of the above equation yields:

$$\lim_{z \to 1}(z-1) X(z) = x(0) + \lim_{z \to 1} \sum_{n=0}^{\infty} [x(n+1) - x(n)] z^{-n}$$
$$= x(0) + [x(1) - x(0)] + [x(2) - x(1)] + [x(3) - x(2)] + \cdots$$
$$= x(\infty)$$

that is, $x(\infty) = \lim_{z \to 1}(z-1) X(z)$.

显然，只有 $X(z)$ 的极点在单位圆内，当 $n \to \infty$ 时 $x(n)$ 才收敛，才可应用终值定理。

Now letting p_1, p_2, $\cdots$, p_N denote the poles of $X(z) = \dfrac{B(z)}{A(z)}$, $x(n)$ has a limit as $n \to \infty$ if and only if the magnitudes $|p_1|$, $|p_2|$, $\cdots$, $|p_N|$ are all strictly less than 1, except that one of the p_i's may be equal to 1. This is equivalent to the condition that all the poles of $(z-1) X(z)$ have magnitudes strictly less than 1. If this condition is satisfied, the limit of $x(n)$ as $n \to \infty$ is given by

$$\lim_{n \to \infty} x(n) = [(z-1) X(z)]_{z=1}$$

Example 6-8: Given a causal sequence $x(n)$ with z-transform $X(z) = \dfrac{z}{z-a}$, $|z| > |a|$, where a is a real, compute the initial-value $x(0)$ and the final-value $x(\infty)$.

Solution: According to Eq. (6-35),

$$x(0) = \lim_{z \to \infty} X(z) = \lim_{z \to \infty} \frac{z}{z-a} = 1$$

when $-1 < a \leq 1$,

$$x(\infty) = \lim_{z \to 1}[(z-1)X(z)]$$

$$= \lim_{z \to 1}(z-1)\frac{z}{z-a} = \begin{cases} 1, & a = 1 \\ 0, & |a| < 1 \end{cases}$$

when $|a| > 1$ or $a = -1$, the final-value theorem isn't applicable, that is, $x(\infty)$ does not exist.

6.3.9 时域卷积定理(Convolution in Time Domain)

If $x(n) \leftrightarrow X(z)$, $ROC = R_1$, $y(n) \leftrightarrow Y(z)$, $ROC = R_2$, then

$$x(n) * y(n) \leftrightarrow X(z)Y(z), \quad ROC = R_1 \cap R_2 \tag{6-37}$$

Proof:

$$\mathscr{Z}[x(n) * y(n)] = \sum_{n=-\infty}^{\infty}\left[\sum_{m=-\infty}^{\infty} x(m)y(n-m)\right]z^{-n}$$

$$= \sum_{m=-\infty}^{\infty} x(m) \cdot \sum_{n=-\infty}^{\infty} y(n-m)z^{-n}$$

$$= \sum_{m=-\infty}^{\infty} x(m)z^{-m} \cdot \sum_{n=-\infty}^{\infty} y(n-m)z^{-(n-m)}$$

$$= X(z) \cdot Y(z)$$

可见，两序列在时域中的卷积等效于在 z 域中两序列 z 变换的乘积。一般情况下，卷积和 $x(n) * y(n)$ 的 z 变换的收敛域是 $X(z)$ 和 $Y(z)$ 收敛域的重叠部分。但若位于某一 z 变换收敛域边缘上的极点被另一 z 变换的零点抵消，则收敛域将会扩大。

Therefore, convolution in the discrete-time domain corresponds to a product in the z-transform domain. This result is obviously analogous to the result in the continuous-time framework where convolution corresponds to multiplication in the s-domain. The ROC of $X(z)Y(z)$ includes the intersection of R_1 and R_2 and may be larger if pole-zero cancellation occurs in the product.

Example 6-9: Given the sequences $x(n) = a^n u(n)$, $h(n) = b^n u(n) - ab^{n-1}u(n-1)$, $|b| < |a|$, calculate the convolution sum $y(n) = x(n) * h(n)$.

Solution: $X(z) = \mathscr{Z}[x(n)] = \dfrac{z}{z-a}, \quad |z| > |a|$

$$H(z) = \mathscr{Z}[h(n)] = \frac{z}{z-b} - az^{-1}\frac{z}{z-b} = \frac{z}{z-b} - \frac{a}{z-b} = \frac{z-a}{z-b}, \quad |z| > |b|$$

then

$$Y(z) = X(z)H(z) = \frac{z}{z-a}\frac{z-a}{z-b} = \frac{z}{z-b}, \quad |z| > |b|$$

therefore,

$$y(n) = x(n) * h(n) = \mathscr{Z}^{-1}[Y(z)] = b^n u(n)$$

Obviously, the ROC of $Y(z)$ is greater than the intersection of $X(z)$ and $H(z)$.

最后，将上面讨论的 z 变换的性质列于表 6-2 中，以便查阅。

表 6-2 z 变换的性质

序号	序列	z 变换	收敛域
1	$ax_1(n)+bx_2(n)$	$aX_1(z)+bX_2(z)$	$R_1 \cap R_2$，若出现零极点抵消，收敛域可能扩大
2	$x(n\pm m)$	$z^{\pm m}X(z)$	R，$z=0$ 或 $z=\infty$ 处可能除外
3	$x(n+m)u(n)$	$z^m\left[X(z)-\sum_{k=0}^{m-1}x(k)z^{-k}\right]$	R，$z=0$ 或 $z=\infty$ 处可能除外
4	$x(n-m)u(n)$	$z^{-m}\left[X(z)+\sum_{k=-m}^{-1}x(k)z^{-k}\right]$	R，$z=0$ 或 $z=\infty$ 处可能除外
5	$a^n x(n)$	$X\left(\dfrac{z}{a}\right)$	$|a|R$
6	$nx(n)$	$-z\dfrac{\mathrm{d}X(z)}{\mathrm{d}z}$	R
7	$\sum_{i=0}^{n}x(i)$	$\dfrac{1}{1-z^{-1}}X(z)$	$|z|>\max(R_x,1)$
8	$x(-n)$	$X\left(\dfrac{1}{z}\right)$	$\dfrac{1}{R}$
9	$x(n)*y(n)$	$X(z)Y(z)$	$R_1 \cap R_2$，若出现零极点抵消，收敛域可能扩大
10	$x(0)=\lim\limits_{z\to\infty}X(z)$		$x(n)$ 为因果序列，$X(z)$ 为真分式
11	$x(\infty)=\lim\limits_{z\to 1}(z-1)X(z)$		$x(n)$ 为因果序列，$X(z)$ 的极点落于单位圆内部，最多在 $z=1$ 处有一极点

6.4 z 反变换（The Inversion z-transform）

在连续时间系统中，应用拉普拉斯变换的目的是把描述连续系统的微分方程转变为复变量 s 的代数方程，然后写出系统的传递函数，即用拉普拉斯反变换法求出系统的时间响应，从而简化了系统的分析。与此类似，在离散系统中应用 z 变换的目的是为了把描述离散系统的差分方程转变为复变量 z 的代数方程，然后写出离散的传递函数，再用 z 反变换求出离散系统的时间响应。求 z 反变换的方法通常有幂级数展开法(长除法)、部分分式展开法及围线积分法(留数法)三种。

Now we turn to the important problem of recovering the sequence $x(n)$ from its z-transform $X(z)$. As we have discussed previously, because we have not restricted the domain of the index set to be either positive or negative integers, we must have knowledge of the region of convergence of $X(z)$ to determine $x(n)$ uniquely. We consider three computation methods of the inverse z-transform—power series expansion, partial fraction expansion and residue methods.

6.4.1 幂级数展开法(Power Series Expansion Method)

由 z 变换的定义可知，$x(n)$ 的 z 变换为 z^{-1} 的幂级数，即

$$X(z)=\sum_{n=0}^{\infty}x(n)z^{-n}=x(0)+x(1)z^{-1}+x(2)z^{-2}+\cdots$$

$$=\frac{B(z)}{A(z)}=\frac{b_m z^m+b_{m-1}z^{m-1}+\cdots+b_0}{a_n z^n+a_{n-1}z^{n-1}+\cdots+a_0} \qquad (6-38)$$

在给定的收敛域内，把 $X(z)$ 展为幂级数，其系数就是序列 $x(n)$。

To compute the inverse z-transform $x(n)$ for a finite range of values of n, $X(z)$ can be expanded into a power series in z^{-1} by dividing $A(z)$ into $B(z)$ using long division. The values of $x(n)$ are coefficients of the power series. The process is illustrated by the following example.

Example 6-10: Suppose that $X(z) = \dfrac{2z^2-0.5z}{z^2-0.5z-0.5}$, $|z|>1$, write out the from of power series by using long division.

Solution: Dividing $A(z)$ into $B(z)$ gives

$$\begin{array}{r}
2 + 0.5z^{-1} + 1.25z^{-2} + \cdots \\
z^2 - 0.5z - 0.5 \overline{)\,2z^2 - 0.5z} \\
\underline{z^2 - z - 1} \\
0.5z + 1 \\
\underline{0.5z - 0.25 - 0.25z^{-1}} \\
1.25 + 0.25z^{-1} \\
\underline{1.25 - 0.625z^{-1} - 0.625z^{-2}} \\
\cdots
\end{array}$$

Power series expansion method is simple and intuitive, whereas it only gets a limited item of $x(n)$. It is difficult to obtain a closed solution of $x(n)$. The search for a more general procedure for finding $x(n)$ from $X(z)$ leads us to the method of partial fraction expansions.

6.4.2 部分分式展开法(Partial Fractions Expansion Method)

在连续时间系统中，用部分分式展开法可以求拉普拉斯反变换，同样在离散时间系统中，当 $X(z)$ 表达式为有理分式时，z 反变换也可以用此法求得。

Let us assume that theorder number of the numerator in Eq. (6-38) is not greater than that of the denominator, that is, $n \geqslant m$.

(1) 互异极点(distinct poles)

设 $X(z)$ 的极点为互异极点，将 $\dfrac{X(z)}{z}$ 进行部分分式展开得：

$$\frac{X(z)}{z} = \sum_{i=0}^{n} \frac{A_i}{z - p_i} \tag{6-39}$$

或

$$X(z) = \sum_{i=0}^{n} \frac{A_i z}{z - p_i} \tag{6-40}$$

其中 p_i 是 $\dfrac{X(z)}{z}$ 的互异极点，系数

$$A_i = \left[(z - p_i)\frac{X(z)}{z}\right]_{z=p_i}, \quad i = 0, 1, \cdots, n \tag{6-41}$$

式(6-40)也可以写为

$$X(z) = A_0 + \sum_{i=1}^{n} \frac{A_i z}{z - p_i}, \quad A_0 = X(z)|_{z=0}$$

若 $X(z)$ 的收敛域为 $|z|>R_x$，则 $x(n)$ 是右边序列，

$$x(n) = A_0\delta(n) + \sum_{i=1}^{n} A_i (p_i)^n u(n)$$

若 $X(z)$ 的收敛域为 $|z|<R_x$，则 $x(n)$ 是左边序列，

$$x(n) = A_0\delta(n) - \sum_{i=1}^{n} A_i (p_i)^n u(-n-1)$$

Example 6-11: Find the inverse z-transform of

$$X(z) = \frac{z^3 - z^2 + z}{\left(z - \frac{1}{2}\right)(z-2)(z-1)} \quad 1 < |z| < 2$$

Solution: We use partial fraction expansions to write

$$\frac{X(z)}{z} = \frac{z^2 - z + 1}{\left(z - \frac{1}{2}\right)(z-2)(z-1)} = \frac{A_1}{z - \frac{1}{2}} + \frac{A_2}{z-2} + \frac{A_3}{z-1}$$

where,

$$A_1 = \frac{z^2 - z + 1}{(z-2)(z-1)}\bigg|_{z=\frac{1}{2}} = \frac{\frac{1}{4} - \frac{1}{2} + 1}{\left(-\frac{3}{2}\right) \times \left(-\frac{1}{2}\right)} = \frac{\frac{3}{4}}{\frac{3}{4}} = 1$$

$$A_2 = \frac{z^2 - z + 1}{\left(z - \frac{1}{2}\right)(z-1)}\bigg|_{z=2} = \frac{4 - 2 + 1}{\frac{3}{2} \times 1} = 2$$

$$A_3 = \frac{z^2 - z + 1}{\left(z - \frac{1}{2}\right)(z-2)}\bigg|_{z=1} = \frac{1 - 1 + 1}{\frac{1}{2} \times (-1)} = -2$$

so that

$$X(z) = \frac{z}{z - \frac{1}{2}} + \frac{2z}{z-2} - \frac{2z}{z-1}$$

Then take the inverse z-transform of each term in above equation. According to the relationship between the location of the poles and ROC of $X(z)$, each of which is depicted in Figure 6-4.

The figure shows that the ROC has a radius greater than the pole at $z=\frac{1}{2}$ and $z=1$, so that these terms have the right-side inverse transform.

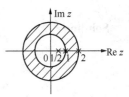

Figure 6-4 ROC of Example 6-11

$$\frac{z}{z - \frac{1}{2}} \xrightarrow{\mathscr{Z}^{-1}} \left(\frac{1}{2}\right)^n u(n)$$

$$-\frac{2z}{z - 1} \xrightarrow{\mathscr{Z}^{-1}} -2u(n)$$

The ROC also has a radius less than the pole at $z = 2$. So this term has the left-side inverse transform

$$\frac{2z}{z-2} \xrightarrow{\mathscr{Z}^{-1}} -2(2)^n u(-n-1)$$

combining the individual terms gives

$$x(n) = \left(\frac{1}{2}\right)^n u(n) - 2(2)^n u(-n-1) - 2u(n)$$

(2) 重极点(repeated poles)

设 p_1 为 r 重极点，而其他极点均为互异的，则可将 $\frac{X(z)}{z}$ 展开为如下形式：

$$\frac{X(z)}{z} = \frac{A_0}{z} + \frac{A_1}{z-p_1} + \frac{A_2}{(z-p_1)^2} + \cdots + \frac{A_r}{(z-p_1)^r} + \frac{B_2}{z-p_2} + \cdots \frac{B_{n-r}}{z-p_{n-r}}$$

即，

$$X(z) = A_0 + \sum_{j=1}^{r} \frac{A_j z}{(z-p_1)^j} + \sum_{i=2}^{n-r} \frac{B_i z}{z-p_i} \qquad (6\text{-}42)$$

其中，

$$A_0 = X(z)\big|_{z=0}$$

$$A_j = \frac{1}{(r-j)!}\left[\frac{\mathrm{d}^{r-j}}{\mathrm{d}z^{r-j}}(z-p_1)^r \frac{X(z)}{z}\right]_{z=p_1}$$

$$B_i = \frac{X(z)}{z}(z-p_i)\big|_{z=p_i}$$

若收敛域为 $|z| > R_x$，则重极点的反变换为右边序列，

$$\frac{A_j z}{(z-p_1)^j} \xrightarrow{\mathscr{Z}^{-1}} A_j \frac{n(n-1)\cdots(n-j+2)}{(j-1)!}(p_1)^{n-j+1} u(n)$$

相反，若收敛域为 $|z| < R_x$，则重极点的反变换为左边序列，

$$\frac{A_j z}{(z-p_1)^j} \xrightarrow{\mathscr{Z}^{-1}} -A_j \frac{n(n-1)(n-2)\cdots(n-j+2)}{(j-1)!}(p_1)^{n-j+1} u(-n-1)$$

Example 6-12: Find the inverse of the z-transform

$$X(z) = \frac{z^3 + 2z^2 - 4}{(z-1)(z+2)^2} \qquad |z| > 2$$

Solution:

$$\frac{X(z)}{z} = \frac{z^3 + 2z^2 - 4}{z(z-1)(z+2)^2} = \frac{A_0}{z} + \frac{A_1}{z+2} + \frac{A_2}{(z+2)^2} + \frac{B_2}{z-1}$$

$$A_0 = \frac{z^3 + 2z^2 - 4}{(z-1)(z+2)^2}\bigg|_{z=0} = \frac{-4}{-4} = 1$$

$$B_2 = \frac{z^3 + 2z^2 - 4}{z(z+2)^2}\bigg|_{z=1} = \frac{1+2-4}{9} = -\frac{1}{9}$$

$$A_1 = \left(\frac{z^3 + 2z^2 - 4}{z(z-1)}\right)\bigg|_{z=-2}$$

$$= \frac{(3z^2 + 4z)(z^2 - z) - (2z-1)(z^3 + 2z^2 - 4)}{(z^2 - z)^2}\bigg|_{z=-2}$$

$$= \frac{(12-8)(4+2) - (-5) \times (-8+8-4)}{(4+2)^2}$$

$$= \frac{4}{36} = \frac{1}{9}$$

$$A_2 = \frac{z^3 + 2z^2 - 4}{z(z-1)}\bigg|_{z=-2} = \frac{-8+8-4}{-2\times(-3)} = \frac{-2}{3}$$

so that
$$X(z) = 1 + \frac{-\frac{1}{9}z}{z-1} + \frac{\frac{1}{9}z}{z+2} + \frac{-\frac{2}{3}z}{(z+2)^2}$$

thus, the inverse z-transform is
$$x(n) = \delta(n) - \frac{1}{9}u(n) + \frac{1}{9}(-2)^n u(n) + \frac{1}{3}n(-2)^n u(n)$$

6.4.3　围线积分法(留数法) (Contour Integral Method (Residue Method))

如本章 6.1 节所述，若已知序列 $x(n)$ 的 z 变换为 $\mathscr{Z}[x(n)] = X(z)$，则 $X(z)$ 的逆变换记作 $\mathscr{Z}^{-1}[X(z)]$，可由以下的围线积分给出：

$$x(n) = \frac{1}{2\pi j}\oint_C X(z) z^{n-1} dz$$

其中 C 是包围 $X(z)z^{n-1}$ 所有极点的逆时针闭合积分路线。

In order to compute the sequence $x(n)$ from the z-transform $X(z)$, we select a closed contour C that takes the counter clock wise direction in the region of convergence.

由于围线 C 在 $X(z)$ 的收敛域内，且包围坐标原点，而 $X(z)$ 又在 $|z|>R$ 的区域内收敛，因此 C 包围了 $X(z)$ 的奇点。通常 $X(z)z^{n-1}$ 是 z 的有理函数，其奇点都是孤立点(极点)。这样，借助复变函数的留数定理，将上述积分表示为围线 C 内所包含 $X(z)z^{n-1}$ 的各极点留数之和，即

$$x(n) = \mathscr{Z}^{-1}[X(z)] = \frac{1}{2\pi j}\oint_C X(z) z^{n-1} dz = \sum_m \operatorname{Re} s[X(z)z^{n-1}]_{z=z_m} \quad (6-43)$$

其中 $\operatorname{Re} s$ 表示极点的留数，z_m 为 $X(z)z^{n-1}$ 的极点。

设 z_m 是 $X(z)z^{n-1}$ 的单一阶极点，则有
$$\operatorname{Re} s[X(z)z^{n-1}]_{z=z_m} = [(z-z_m)X(z)z^{n-1}]_{z=z_m}$$

若 z_m 是 $X(z)z^{n-1}$ 的多重(r 重)极点，则有

$$\operatorname{Re} s[X(z)z^{n-1}]_{z=z_m} = \frac{1}{(r-1)!}\frac{d^{r-1}}{dz^{r-1}}[(z-z_m)^r X(z)z^{n-1}]_{z=z_m} \quad (6-44)$$

When using the above formulas, we must note the order of the poles. Those are inside the contour C. For different n, the poles $z=0$ may have different order.

Example 6-13: Determine the unilateral inverse z-transform of

$$X(z) = \frac{z^2}{(z-0.2)(z+0.8)}, \qquad |z|>0.8$$

Solution:
$$x(n) = \frac{1}{2\pi j}\oint_C X(z)z^{n-1}dz = \sum_i \operatorname{Re} s[X(z)z^{n-1}]_{z_i} = \sum_i \operatorname{Re} s\left[\frac{z^{n+1}}{(z-0.2)(z+0.8)}\right]_{z_i}$$

for $n\geq 0$, there are two poles inside C, $z=0.2$ and $z=-0.8$. Therefore we have that

$$x(n) = \left.\frac{z^{n+1}}{z+0.8}\right|_{z=0.2} + \left.\frac{z^{n+1}}{z-0.2}\right|_{z=-0.8}$$

$$= (0.2)^{n+1} - (-0.8)^{n+1}$$

that is,
$$x(n) = [(0.2)^{n+1} - (-0.8)^{n+1}]u(n)$$

6.5 利用 z 变换求解差分方程(Solving Difference Equation by Using z-transform)

描述离散时间系统的差分方程可通过 z 变换转换为代数方程求解,由于一般的激励及响应都是有始序列,所以下面只讨论单边 z 变换求解差分方程的问题。

Discrete-time systems described by difference equations can be transformed into algebraic equations to solve through z-transform. The following discussion is only the problem of unilateral z-transform to solve difference equations.

6.5.1 零输入响应的 z 域求解(The z-domain Solution of Zero Input Response)

对于线性时不变离散时间系统,在零输入条件下,即激励 $x(n)=0$ 时,其差分方程为

$$\sum_{i=0}^{k} a_i y(n-i) = 0 \tag{6-45}$$

考虑响应为 $n \geq 0$ 时的值,则初始条件为 $y(-1)$,$y(-2)$,…,$y(-n)$。将式(6-45)两边取单边 z 变换,并根据 z 变换的位移性质,可得

$$\sum_{i=0}^{k} a_i z^{-i} \left[Y(z) + \sum_{n=-i}^{-1} y(n) z^{-n} \right] = 0$$

所以

$$Y(z) = \frac{-\sum_{i=0}^{k} \left[a_i z^{-i} \cdot \sum_{n=-i}^{-1} y(n) z^{-n} \right]}{\sum_{i=0}^{k} a_i z^{-i}} \tag{6-46}$$

响应的序列可由 z 反变换求得。

Example 6-14: Consider an LTI system described by the difference equation
$$y(n) - 5y(n-1) + 6y(n-2) = x(n)$$
the initial conditions are $y(-2)=1$,$y(-1)=4$,find the zero input response $y_{zi}(n)$.

Solution: When the input is zero, that is $x(n)=0$, then
$$y(n) - 5y(n-1) + 6y(n-2) = 0$$
taking the unilateral z-transform on both sides, we have
$$Y(z) - 5z^{-1}[Y(z) + y(-1)z] + 6z^{-2}[Y(z) + y(-1)z + y(-2)z^2] = 0$$

that is
$$Y(z) = \frac{5y(-1) - 6z^{-1}y(-1) - 6y(-2)}{1 - 5z^{-1} + 6z^{-2}} = \frac{14z^2 - 24z}{z^2 - 5z + 6}$$

$$\frac{Y(z)}{z} = \frac{14z-24}{(z-2)(z-3)} = \frac{-4}{z-2} + \frac{18}{z-3}$$

therefore,
$$Y(z) = \frac{-4z}{z-2} + \frac{18z}{z-3}$$

so that the zero input response is
$$y_{zi}(n) = [18(3)^n - 4(2)^n],\ n \geq 0$$

6.5.2 零状态响应的 z 域求解(The z-domain Solution of Zero State Response)

N 阶线性时不变离散时间系统的差分方程为

$$\sum_{i=0}^{N} a_i y(n-i) = \sum_{j=0}^{M} b_j x(n-j) \tag{6-47}$$

在零状态条件下,即 $y(-1) = y(-2) = \cdots = y(-n) = 0$ 时,将等式(6-46)两边取单边 z 变换可得

$$\sum_{i=0}^{N} a_i z^{-i} Y(z) = \sum_{j=0}^{M} b_j z^{-j} X(z)$$

其中激励序列 $x(n)$ 为因果序列,即当 $n<0$ 时, $x(n) = 0$,且 $M \leq N$,则有

$$Y(z) = X(z) \frac{\sum_{r=0}^{m} b_r z^{-r}}{\sum_{i=0}^{k} a_i z^{-i}} \tag{6-48}$$

零状态响应的序列可由 z 反变换求得。

Example 6-15: Consider an LTI system described by the difference equation
$$y(n) - 5y(n-1) + 6y(n-2) = x(n)$$
where, $x(n) = 4^n u(n)$ and $y(-2) = y(-1) = 0$, find the zero state response $y_{zs}(n)$.

Solution: Because
$$X(z) = \mathscr{Z}[x(n)] = \frac{z}{z-4}$$

according to Eq. (6-48) we have
$$Y(z) = \frac{z}{z-4} \cdot \frac{1}{1 - 5z^{-1} + 6z^{-2}} = \frac{z^3}{(z-4)(z^2 - 5z + 6)}$$

$$\frac{Y(z)}{z} = \frac{z^2}{(z-4)(z^2 - 5z + 6)} = \frac{2}{z-2} - \frac{9}{z-3} + \frac{8}{z-4}$$

that is,
$$Y(z) = \frac{2z}{z-2} - \frac{9z}{z-3} + \frac{8z}{z-4}$$

therefore, the zero state response is
$$y_{zs}(n) = [2 \cdot 2^n - 9 \cdot 3^n + 8 \cdot 4^n] u(n)$$

6.5.3 全响应的 z 域求解(The z-domain Solution of Complete Response)

对于线性时不变离散时间系统,若激励和初始状态均不为零,则对应的响应称之为全响应,可按下式计算:

$$y(n) = y_{zi}(n) + y_{zs}(n)$$

The method for solving $y_{zi}(n)$ and $y_{zs}(n)$ had been mentioned in above examples. In addition, the complete response $y(n)$ can also be computed by taking the unilateral z-transform on both sides of the Eq. (6-47), that is,

$$\sum_{i=0}^{N} a_i z^{-i} \left[Y(z) + \sum_{n=-i}^{-1} y(n) z^{-n} \right] = \sum_{j=0}^{M} b_j z^{-j} \left[X(z) + \sum_{k=-j}^{-1} x(k) z^{-k} \right] \quad (6-49)$$

Example 6-16: Consider an LTI system described by the difference equation

$$y(n) - 5y(n-1) + 6y(n-2) = x(n)$$

where, $x(n) = 4^n u(n)$ and $y(-2) = 1$, $y(-1) = 4$, find the complete response $y(n)$.

Solution: Because

$$X(z) = \mathscr{Z}[x(n)] = \frac{z}{z-4}$$

taking the unilateral z-transform on both sides of the difference equation:

$$Y(z) - 5z^{-1}Y(z) - 5y(-1) + 6z^{-2}Y(z) + 6y(-1)z^{-1} + 6y(-2) = X(z)$$

yields,

$$Y(z) = \frac{X(z) + 5y(-1) - 6y(-2) - 6y(-1)z^{-1}}{1 - 5z^{-1} + 6z^{-2}}$$

$$= \frac{15z^3 - 80z^2 + 96z}{(z-4)(z^2 - 5z + 6)}$$

$$\frac{Y(z)}{z} = \frac{15z^2 - 80z + 96}{(z-4)(z-2)(z-3)} = \frac{8}{z-4} + \frac{9}{z-3} - \frac{2}{z-2}$$

that is,

$$Y(z) = \frac{8z}{z-4} + \frac{9z}{z-3} - \frac{2z}{z-2}$$

therefore, the complete response is

$$y(n) = [8 \cdot 4^n + 9 \cdot 3^n - 2 \cdot 2^n] u(n)$$

Obviously, the result is the same as the summation of Example 6-14 and Example 6-15.

6.6 离散时间系统的系统函数与频率响应(System Function and Frequency Response of Discrete-time System)

6.6.1 系统函数的定义(Definition of the System Function)

一个线性时不变离散时间系统在时域中可以用它的单位脉冲响应 $h(n)$ 来表示,则零状态响应为 $y(n) = x(n) * h(n)$。

对等式两边做 z 变换,由时域卷积定理得

$$Y(z) = X(z) \cdot H(z)$$

则

$$H(z) = \frac{Y(z)}{X(z)} \quad (6-50)$$

$H(z)$ 称为线性时不变离散时间系统的系统函数,它是单位脉冲响应的 z 变换,即

$$H(z) \triangleq \mathscr{Z}[h(n)] = \sum_{n=-\infty}^{\infty} h(n) z^{-n}$$

This definition implies that the transfer function may also be viewed as the ratio of the z-transforms of the output and input. Note this definition applies at all z in the ROC of $X(z)$ and $Y(z)$, and $X(z)$ is nonzero.

系统函数可以直接从描述离散系统的差分方程得到。设 N 阶差分方程为

$$\sum_{i=0}^{N} a_i y(n-i) = \sum_{j=0}^{M} b_j x(n-j)$$

若激励信号 $x(n)$ 为因果信号，且系统处于零状态，对上式两边取 z 变换，可得

$$\sum_{i=0}^{N} a_i z^{-i} Y(z) = \sum_{j=0}^{M} b_j z^{-j} X(z)$$

所以，

$$H(z) = \frac{\sum_{j=0}^{M} b_j z^{-j}}{\sum_{i=0}^{N} a_i z^{-i}} \tag{6-51}$$

From the above we can see, the coefficients of numerator and denominator polynomial of the system function correspond to the coefficients of the difference equation. This correspondence allows us not only to find the system function through giving the difference equation, but also to find a difference equation through giving a rational system function.

Example 6-17: Determine the system function for given system described by the difference equation

$$y(n) - 3y(n-1) + 2y(n-2) = 2x(n)$$

Solution: From Eq. (6-51), we obtain the system function

$$H(z) = \frac{2}{1 - 3z^{-1} + 2z^{-2}} \tag{6-52}$$

Similarly, we can obtain the difference equation from the system function

$$H(z) = \frac{2}{1 - 3z^{-1} + 2z^{-2}} = \frac{Y(z)}{X(z)}$$

that is,

$$Y(z) - 3z^{-1} Y(z) + 2z^{-2} Y(z) = 2X(z)$$

From Eq. (6-52) with $H(z)$ viewed as a ratio of polynomials in z, the order of the numerator does not exceed that of the denominator, and thus we can conclude that the LTI system is causal. And we can write the difference equation that, together with the condition of initial rest, characterizes the system

$$y(n) - 3y(n-1) + 2y(n-2) = 2x(n)$$

6.6.2 系统函数的零极点分布(Zeros-Poles Distribution of the System Function)

将式(6-51)进行因式分解，可得

$$H(z) = H_0 \frac{\prod_{j=1}^{M}(1 - z_j z^{-1})}{\prod_{i=1}^{N}(1 - p_i z^{-1})} \tag{6-53}$$

式中，z_j 和 p_i 分别是 $H(z)$ 的零点和极点，$H_0 = \dfrac{b_0}{a_0}$ 是 $H(z)$ 的增益系数。可见，除了 H_0 之外，系统函数完全由它的零极点来确定。

Zeros-poles graph also describes the system function, in general, "○" represents the zero and "×" represents the pole in zeros-poles graph. For example, system function

$$H(z) = \frac{z}{(z-1)(z+1)}$$

then, zeros-poles distribution of the system function is shown in Figure 6-5.

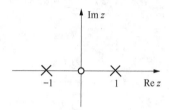

Figure 6-5 zeros-poles distribution graph

将式(6-53)进行部分分式展开，则 $H(z)$ 还可以表示为

$$H(z) = \sum_{i=1}^{N} \frac{A_i}{1 - p_i z^{-1}} \tag{6-54}$$

对每个部分分式取 z 反变换即可得系统的单位脉冲响应 $h(n)$。显然，$H(z)$ 的极点决定了 $h(n)$ 的波形；而 $H(z)$ 的零点决定了 $h(n)$ 的幅度和相位特性。

From Eq. (6-54), it follows that the waveform of the unit impulse response $h(n)$ is directly determined by the poles of the system function $H(z)$. While, the magnitude and phase of $h(n)$ are affected by the zeros of the system function. Therefore, we can determine the nature of unit impulse response $h(n)$ from the zeros-poles distribution of $H(z)$.

下面讨论 $H(z)$ 的极点与 $h(n)$ 波形的关系。

(1) 单实数极点 $p = r$

$$h(n) = r^n u(n)$$

若 $r>1$，极点在单位圆外，$h(n)$ 为增幅指数序列；若 $r<1$，极点在单位圆内，$h(n)$ 为衰减指数序列；若 $r=1$，极点在单位圆上，$h(n)$ 为等幅序列。

(2) 共轭极点 $p_{1,2} = re^{\pm j\theta}$

$$H(z) = \frac{A}{1 - re^{j\theta}z^{-1}} + \frac{A^*}{1 - re^{-j\theta}z^{-1}}$$

为分析方便起见，令 $A=1$，可得对应系统的单位脉冲响应为

$$h(n) = 2r^n \cos(\theta n) u(n)$$

若 $r>1$，极点在单位圆外，$h(n)$ 为增幅振荡序列；若 $r<1$，极点在单位圆内，$h(n)$ 为衰减振荡序列；若 $r=1$，极点在单位圆上，$h(n)$ 为等幅振荡序列。

Figure 6-6 depicts the relationship between the location of poles and the waveform of $h(n)$.

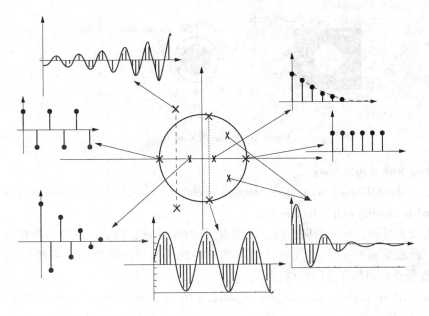

Figure 6-6 the relationship between the location of poles and the waveform of $h(n)$

6.6.3 因果稳定系统(Causal and Stable Discrete-time System)

(1) 因果性(causality)

时域判断离散时间 LTI 系统因果的充要条件是单位脉冲响应 $h(n)$ 为因果序列。

The impulse response $h(n)$ of a causal LTI system is zero for $n<0$, and therefore it is a right-sided sequence.

z 域等效条件是 $H(z)$ 的收敛域必须是某个圆的圆外区域，且必须包含 $z=\infty$ 在内。

A discrete-time LTI system is causal if and only if the ROC of $H(z)$ is the exterior of a circle in the z-plane, including infinity.

A discrete-time LTI system with rational system function $H(z)$ is causal if and only if:

(a) The ROC is the exterior of a circle outside the outermost pole.

(b) With $H(z)$ expressed as a ratio of polynomials in z, the order of the numerator cannot be greater than the order of the denominator.

(2) 稳定性(stability)

时域判断离散时间 LTI 系统稳定的充要条件是单位脉冲响应绝对可和，即

$$\sum_{n=-\infty}^{\infty} |h(n)| < \infty \qquad (6-55)$$

The stability of a discrete-time LTI system is equivalent to its impulse response being absolutely summable.

上式的 z 域等效条件是 $H(z)$ 的收敛域包括单位圆。

It also follows that the ROC of $H(z)$ must include the unit circle in the z-plane. This can be explained as follow. Let a locates in the inner of unit circle, as illustrated in Figure 6-7(a) and Figure 6-7(b).

For (a), the ROC include the unit circle, so that $h(n) = a^n u(n)$ is a right-sided sequence

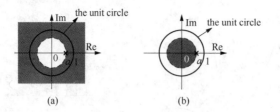

Figure 6-7 the ROC of $H(z)$

and decreasing with n increasing.

For (b), the ROC does not include the unit circle, so that $h(n) = -a^n u(-n-1)$ is a left-sided sequence and increasing with n increasing.

由以上分析可知，对于因果系统，要求 $H(z)$ 的收敛域必须是一个以最外极点为半径的圆外区域，而稳定系统要求收敛域必须包括单位圆。因此，因果稳定系统的充要条件是 $H(z)$ 的所有极点必须都在单位圆内。

A causal LTI system with rational system function $H(z)$ is stable if and only if all of the poles of $H(z)$ lie inside the unit circle—i.e., they must all have magnitude smaller than 1.

Example 6-18: Consider a system with system function

$$H(z) = \frac{\left(a - \frac{1}{a}\right)z}{(z-a)\left(z-\frac{1}{a}\right)} = \frac{z}{z-a} - \frac{z}{z-\frac{1}{a}}, \text{ where } 0 < |a| < 1,$$

analysis the causality and stability of the system.

Solution: the poles of this system function are $z = a$, $z = \frac{1}{a}$.

① When the ROC is $\frac{1}{a} < |z| \leq \infty$, we can conclude that the system is **causal**, but the ROC doesn't include the unit circle, and consequently we can confirm that the system is **not stable**. The impulse response of this system is $h(n) = \left[a^n - \left(\frac{1}{a}\right)^n\right]u(n)$, which is only a causal sequence.

② When the ROC is $0 \leq |z| \leq a$, we can conclude that the system is **non-causal**, and the ROC doesn't include the unit circle, so that the system is **not stable**. The impulse response of this system is $\left[\left(\frac{1}{a}\right)^n - a^n\right]u(-n-1)$, which is a non-causal sequence.

③ When the ROC is $a < |z| \leq \frac{1}{a}$, the system is **non-causal**, and the ROC includes the unit circle, so that the system is **stable**. The impulse response of this system is $a^n u(n) + \left(\frac{1}{a}\right)^n u(-n-1)$, which is a bilateral sequence.

6.6.4 离散系统的频率响应(Frequency Response of the Discrete-time System)

在连续时间系统中，系统的频率响应特性反映了系统在正弦激励下的稳态响应随频率变化的情况，与此相似的是，在离散系统中，也要研究正弦序列激励下的稳态响应随频率变化的关系，即离散系统的频率响应特性及其意义。

As in the continuous-time case, the frequency response characteristics of the system can be determined by examining the response to a sinusoidal input. Similarly, in discrete-time systems, the relationship between the steady state response to sinusoidal sequence and the frequency should be also researched. Throughout this section it is assume that the LTI discrete-time system is stable.

设输入序列是频率为 ω 的复指数序列，即
$$x(n) = e^{jn\omega}, \quad -\infty < n < \infty$$
利用卷积和可得
$$\begin{aligned}
y(n) = h(n) * x(n) &= \sum_{k=0}^{\infty} h(k) x(n-k) \\
&= \sum_{k=0}^{\infty} h(k) e^{-jk\omega} e^{jn\omega} \\
&= e^{jn\omega} \sum_{k=0}^{\infty} h(k) (e^{j\omega})^{-k} \\
&= e^{jn\omega} H(z)|_{z=e^{j\omega}} \\
&= e^{jn\omega} H(e^{j\omega})
\end{aligned} \qquad (6-56)$$

Eq. (6-56) implied that the steady state response for complex exponential sequence is always of the form $H(e^{j\omega}) e^{jn\omega}$. In other words, the steady state response is identical exponential, modified in amplitude and phase by system function $H(e^{j\omega})$.

$$H(e^{j\omega}) \triangleq H(z)|_{z=e^{j\omega}} = |H(e^{j\omega})| e^{j\varphi(\omega)} \qquad (6-57)$$

$H(e^{j\omega})$ 称为 LTI 离散时间系统的频率响应，它描述了复指数序列通过 LTI 系统后复振幅的变化。$|H(e^{j\omega})|$ 称为系统的幅度响应，$\varphi(\omega)$ 称为系统的相位响应。

Where, $H(e^{j\omega})$ is called frequency response, $|H(e^{j\omega})|$ is called amplitude response and $\varphi(\omega)$ is called phase response of the system.

由式(6-57)可以看出，系统函数 $H(z)$ 在 z 平面中令 z 沿单位圆变化即得系统的频率响应 $H(e^{j\omega})$。可见，离散系统的频率响应是以 2π 为周期的周期谱。

The frequency response of the discrete-time system is just the value of $H(z)$ around the unit circle in the z-plane. Consequently, the frequency response $H(e^{j\omega})$ is periodic, and the period is 2π.

Example 6-19: Find the system function, impulse response and the frequency response. The difference equation of system is given by
$$y(n) = ay(n-1) + x(n) \quad |a| < 1$$

Solution: The system function can easily obtained by z-transform:
$$Y(z) = aY(z)z^{-1} + X(z)$$
then
$$H(z) = \frac{Y(z)}{X(z)} = \frac{1}{1-az^{-1}} \quad |z| > a$$

this is a causal system, so the impulse response is
$$h(n) = a^n u(n)$$
according to the Eq. (6-57), we get the frequency response
$$H(e^{j\omega}) = \frac{1}{1 - ae^{-j\omega}} = \frac{1}{1 - a\cos\omega + ja\sin\omega}$$

the amplitude response is
$$|H(e^{j\omega})| = \frac{1}{\sqrt{1+a^2-2a\cos\omega}}$$

and the phase response is
$$\varphi(\omega) = -\arctan\frac{a\sin\omega}{1-a\cos\omega}$$

6.7 利用 MATLAB 进行离散系统的复频域分析(Analysis of Discrete-time Systems in z-Domain Based on MATLAB)

6.7.1 部分分式展开的 MATLAB 实现(MATLAB Realization of Partial Fractions Expansion)

MATLAB 的信号处理工具箱提供了一个对 $X(z)$ 进行部分分式展开的函数 residuez,它的调用格式如下:

$$[r, p, k] = \text{residuez}(\text{num}, \text{den})$$

其中,**num**,**den** 分别表示 $X(z)$ 的分子和分母多项式的系数向量,r 为部分分式的系数,p 为极点,k 为多项式的系数。若 $X(z)$ 为真分式,则 k 为空。借助 residuez 函数可以将 $X(z)$ 展开成

$$X(z) = \frac{num(z)}{den(z)} = \frac{r(1)}{1-p(1)z^{-1}} + \cdots + \frac{r(n)}{1-p(n)z^{-1}} + k(1) + k(2)z^{-1} + \cdots$$

Example 6-20: Expand $X(z)$ to the sum of partial fractions.

$$X(z) = \frac{1+4z^{-3}}{1+2z^{-1}+3z^{-2}+2z^{-3}}$$

Solution: Run the following MATLAB commands:

num = [1 0 0 4];
den = [1 2 3 2];
[r, p, k] = residuez(num, den)

The running results are:

r =
   0.2500 + 1.2284i
   0.2500 - 1.2284i
  -1.5000

p =
  -0.5000 + 1.3229i
  -0.5000 - 1.3229i
  -1.0000

k = 2

Therefore, $X(z)$ can be expanded to the following form:

$$X(z) = \frac{0.25+1.2284i}{1-(-0.5+1.3229i)z^{-1}} + \frac{0.25-1.2284i}{1-(-0.5-1.3229i)z^{-1}} + \frac{-1.5}{1+z^{-1}} + 2$$

6.7.2 基于MATLAB的离散时间系统z域分析(MATLAB Analysis of Discrete-time Systems in z-Domain)

若系统函数 $H(z)$ 为分子分母多项式的形式，则可以借助函数 tf2zp( ) 得到零极点形式的表述式，tf2zp( ) 的调用形式为

$$[z, p, k] = \text{tf2zp}(\text{num}, \text{den})$$

式中 ***num*** 和 ***den*** 分别为分子和分母多项式的系数向量。

Example 6-21: The system function of a causal discrete-time system is given as follow:

$$H(z) = \frac{1 - 3z^{-1} + 3z^{-2} - z^{-3}}{1 + 0.368z^{-1} + 0.5z^{-2} + 0.732z^{-3}}$$

compute the zeros and poles of the system.

Solution: Run the following MATLAB commands:
num = [1 -3 3 -1];
den = [10.368 0.5 0.732];
[r, p, k] = tf2zp(num, den)
run result is:
r =
   1.0000
   1.0000 + 0.0000i
   1.0000 - 0.0000i
p =
   0.2307 + 0.9107i
   0.2307 - 0.9107i
   -0.8293
k = 1

We can also apply the special function zplane( ) to draw the zero-pole distribution graph. Its call format is zplane(num, den). In case of the above example, we run the command
zplane(num, den)

The zero-pole graph is shown in Figure 6-8.

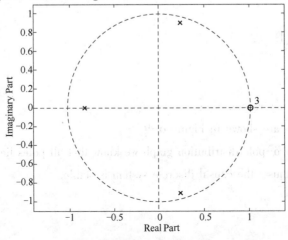

Figure 6-8　zero-pole graph

We may compute the unit impulse response and the frequency response by using the special function impz( ) and freqz( ). The calls format of them will be stated in the following example.

Example 6-22: The system function of a causal discrete-time system is given as follow:

$$H(z) = \frac{1 - 3z^{-1} + 3z^{-2} - z^{-3}}{1 + 0.162z^{-1} + 0.34z^{-2} + 0.15z^{-3}}$$

draw the zero-pole distribution graph of the system, compute the unit impulse response $h(n)$ and the frequency response $H(e^{jw})$, and determine whether the system is stable.

Solution: Run the following MATLAB commands:

num=[1 -3 3 -1];
den=[10.162 0.34 0.15];
 figure(1);
 zplane(num, den)
 h=impz(num, den, 21);
 figure(2);
 stem(0:20, h);
 xlabel('n');
 ylabel('h(n)');
title('unit impulse response');
 [H, w]=freqz(num, den);
 figure(3);
 subplot(211)
plot(w/pi, abs(H));
xlabel('w');
 ylabel('abs(H)')
title('magnitude frequency response');
subplot(212)
plot(w/pi, angle(H));
xlabel('w');
ylabel('angle(H)')
title('phase frequency response');

The running results are shown in Figure 6-9.

According to the zero-pole distribution graph we know that all poles lie in the inner of the unit circle in the z-plane, thus, the causal discrete system is stable.

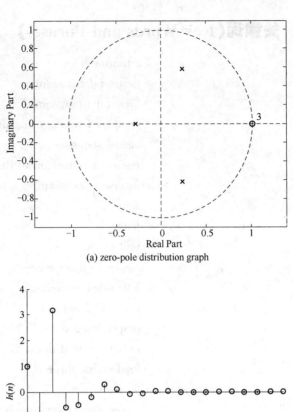

(a) zero-pole distribution graph

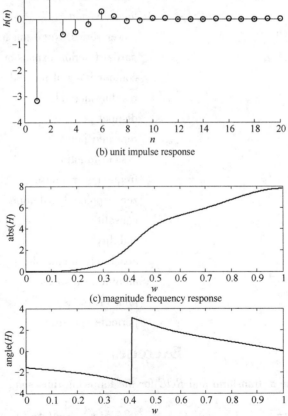

(b) unit impulse response

(c) magnitude frequency response

(d) phase frequency response

Figure 6-9 the waveform of the Example 6-21

关键词(Key Words and Phrases)

(1) z 变换　　　　　　　　　　z transform
(2) 单边 z 变换　　　　　　　　unilateral z transform
(3) 双边 z 变换　　　　　　　　bilateral z transform
(4) 复频域　　　　　　　　　　complex frequency domain
(5) 因果序列　　　　　　　　　causal sequence
(6) 收敛域　　　　　　　　　　region of convergence(ROC)
(7) z 反变换　　　　　　　　　inverse z transform
(8) 零点　　　　　　　　　　　zeros
(9) 极点　　　　　　　　　　　poles
(10) 单位圆　　　　　　　　　　unit circle
(11) 右边序列　　　　　　　　　right-sided sequence
(12) 左边序列　　　　　　　　　left-sided sequence
(13) 双边序列　　　　　　　　　two-sided sequence
(14) 真分式　　　　　　　　　　proper fraction
(15) 初值定理　　　　　　　　　initial-value theorem
(16) 终值定理　　　　　　　　　final-value theorem
(17) 长除法　　　　　　　　　　long division
(18) 幂级数展开法　　　　　　　power series expansion method
(19) 部分分式展开法　　　　　　partial fraction expansion method
(20) 围线积分法　　　　　　　　contour integral method
(21) 留数法　　　　　　　　　　residue method
(22) 互异极点　　　　　　　　　distinct poles
(23) 重极点　　　　　　　　　　repeated poles
(24) 系统函数　　　　　　　　　system function
(25) 频率响应　　　　　　　　　frequency response
(26) 零极点分布　　　　　　　　zeros-poles distribution
(27) 因果性　　　　　　　　　　causality
(28) 稳定性　　　　　　　　　　stability
(29) 绝对可和　　　　　　　　　absolutely summable
(30) 幅度响应　　　　　　　　　amplitude response
(31) 相位响应　　　　　　　　　phase response
(32) 周期谱　　　　　　　　　　periodic spectrum

Exercises

6-1 Determine the z-transform and ROC for the following time signals.

(1) $3\delta(n-2)+\delta(n-5)$　　　　　　(2) $\delta(n)-\dfrac{1}{8}\delta(n-3)$

(3) $\left(\dfrac{1}{2}\right)^n u(n)$　　　　　　　　　(4) $\left(\dfrac{1}{2}\right)^n [u(n)-u(n-10)]$

(5) $n^2 u(n)$ (6) $a^n u(n-N)$

6-2 Let $\mathscr{Z}[x(n)] = X(z) = \dfrac{1}{1+z^{-2}}$, $|z|>1$, using the properties of the z-transform, determine the unilateral z-transform and ROC of the following sequence:

(1) $x_1(n) = x(n-2)$ (2) $x_2(n) = \left(\dfrac{1}{2}\right)^n x(n)$

(3) $x_3(n) = \left(\dfrac{1}{2}\right)^n x(n-2)$ (4) $x_4(n) = nx(n)$

6-3 According to the z-transform $X(z)$, calculate the initial value and the final value of the causal sequence $x(n)$.

(1) $X(z) = \dfrac{z(z+1)}{(z^2-1)(z+0.5)}$ (2) $X(z) = \dfrac{2z^2}{\left(z-\dfrac{1}{2}\right)\left(z+\dfrac{1}{3}\right)}$

6-4 Determine the inverse z-transform for given $X(z)$.

(1) $\dfrac{1}{1+0.5z^{-1}}$ $|z|>0.5$ (2) $\dfrac{1-\dfrac{1}{2}z^{-1}}{1+\dfrac{3}{4}z^{-1}+\dfrac{1}{8}z^{-2}}$ $|z|>\dfrac{1}{2}$

(3) $\dfrac{z-a}{1-az}$ $|z|>\left|\dfrac{1}{a}\right|$ (4) $\dfrac{z^2}{z^2+3z+2}$ $|z|>2$

(5) $\dfrac{z^2+z+1}{z^2+3z+2}$ $|z|>2$ (6) $\dfrac{z^2-az}{(z-a)^3}$ $|z|>|a|$

6-5 Find the inverse z-transform by using power series expansion, partial fractions expansion and residue method.

$$X(z) = \dfrac{10z}{(z-1)(z-2)} \quad |z|>2$$

6-6 Given z-transform $X(z) = \dfrac{2z+1}{z^2+3z+2}$, determine the time sequence $x(n)$ for given three of ROC.

(1) $|z|>2$ (2) $|z|<1$ (3) $1<|z|<2$

6-7 Consider the rectangular signal

$$x(n) = \begin{cases} 1, & 0 \leqslant n \leqslant 5 \\ 0, & otherwise \end{cases}$$

Let $g(n) = x(n) - x(n-1)$,

(1) Find the signal $g(n)$ and directly evaluate its z-transform.

(2) Noting that $x(n) = \sum\limits_{k=-\infty}^{n} g(k)$, determine the z-transform of $x(n)$.

6-8 The difference equation of the causal system is $y(n) + 3y(n-1) = x(n)$.

(1) Find the impulse response of the system.

(2) If $x(n) = (n+n^2)u(n)$, find the complete response $y(n)$, where $y(-1) = 0$.

6-9 A discrete-time system is described by the following difference equation:

$$y(n) + y(n-2) = 2x(n-1) - x(n-2)$$

(1) Compute the unit impulse response $h(n)$.

(2) Compute $y(n)$ for all $n \geq 0$ when $x(n) = 2^n u(n)$ with $y(-1) = 3$ and $y(-2) = 2$.

6-10 A discrete-time system is given by the following difference equation:
$$y(n) + y(n-1) - 2y(n-2) = 2x(n) - x(n-1)$$
Find the input $x(n)$ with $x(n) = 0$ for $n < 0$ that gives the output response $y(n) = 2[u(n) - u(n-3)]$ with initial conditions $y(-2) = 2$, $y(-1) = 0$.

6-11 Determine the system function $H(z)$ and the impulse response $h(n)$, the difference equation are given as following.

(1) $y(n) - 2y(n-1) = x(n)$

(2) $y(n) - 3y(n-1) + 2y(n-2) = x(n)$

(3) $y(n) - 5y(n-1) + 6y(n-2) = x(n) - 3x(n-2)$

6-12 The input $x(n) = (0.5)^n u(n)$ is applied to an LTI discrete-time system with the initial conditions $y(-2) = 4$, $y(-1) = 8$. The resulting output response is
$$y(n) = 4(0.5)^n u(n) - n(0.5)^n u(n) - (-0.5)^n u(n),$$
Find the transfer function $H(z)$.

6-13 A discrete-time system is given by the following difference equation:
$$y(n) - \frac{3}{4}y(n-1) + \frac{1}{8}y(n-2) = x(n)$$
Find the system function $H(z)$, the unit impulse response $h(n)$ and the unit step response $g(n)$.

6-14 A discrete-time LTI system with no initial energy, when the input $x(n) = u(n)$, the output response $y(n) = \left[\left(\frac{1}{2}\right)^n - \left(\frac{1}{3}\right)^n + 2\right]u(n)$, determine the difference equation of the system.

6-15 A causal discrete-time LTI system has the system function $H(z) = \dfrac{1-z^{-1}}{6+5z^{-1}+z^{-2}}$, find:

(1) The unit impulse response $h(n)$.

(2) The difference equation of the system.

(3) Plot the imitation graph of the system.

(4) Determine the stability of the system.

6-16 The zeros-poles graph for a discrete-time LTI system is shown in Figure 6-10, the initial conditions are $y(-1) = 2$, $y(-2) = 1$, note the $\lim_{n \to \infty} h(n) = \dfrac{1}{3}$, find:

(1) The system function $H(z)$ and the difference equation for this system.

(2) The zero input response $y_{zi}(n)$.

(3) If the input is $(-3)^n u(n)$, find the zero state response $y_{zs}(n)$.

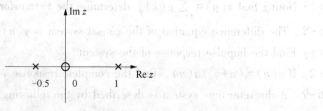

Figure 6-10 the zero-poles graph for Example 6-16

6-17 If the input of a discrete-time system is $x(n) = u(n)$ and the zero state response is $y_{zs}(n) = 2(1-0.5n)u(n)$, then determine the zero state response $y_{zs}(n)$ when input is $x(n) = (0.5)^n u(n)$.

6-18 If the impulse response of a discrete-time system $h(n) = \left(\frac{1}{3}\right)^n u(n) - \left(\frac{1}{4}\right)^n u(n)$, find:

(1) The system function $H(z)$ of the system.

(2) The difference equation of the system.

(3) Draw the imitation graph of the discrete-time system.

(4) Judge the stability and causality of the system.

6-19 For each of the following difference equations and associated input and initial conditions, determine the zero input and zero state responses by using the unilateral z-transform:

(1) $y(n) + 3y(n-1) = x(n)$, $x(n) = \left(\frac{1}{2}\right)^n u(n)$, $y(-1) = 1$

(2) $y(n) - \frac{1}{2}y(n-1) = x(n) - \frac{1}{2}x(n-1)$, $x(n) = u(n)$, $y(-1) = 0$

(3) $y(n) - \frac{1}{2}y(n-1) = x(n) - \frac{1}{2}x(n-1)$, $x(n) = u(n)$, $y(-1) = 1$

6-20 A causal LTI system is described by the difference equation
$$y(n) = y(n-1) + y(n-2) + x(n-1)$$

(1) Find the system function $H(z)$ for this system. Plot the poles and zeros of $H(z)$ and indicate the ROC.

(2) Find the unit impulse response of the system.

(3) You should have found the system to be unstable. Find a stable (noncausal) unit impulse response that satisfies the difference equation.

参 考 文 献 （Reference）

[1] 王明泉. 信号与系统[M]. 北京：科学出版社，2008.

[2] 陈后金，胡健，薛健. 信号与系统[M]. 北京：高等教育出版社，2007.

[3] 郑君里，应启珩，杨为理. 信号与系统引论[M]. 北京：高等教育出版社，2009.

[4] Alan V. Oppenheim. Signals and Systems[M]. 2nd ed. Beijing：Electronic Industry Press，2009.

[5] Edward W. Kamen. Fundamentals of Signals and Systems[M]. 3rd ed. Beijing：Electronic Industry Press，2007.

[6] B. P. Lathi. Linear Systems and Signals[M]. Oxford：Oxford University Press，2005.

[7] S. S. Haykin. Signals and Systems[M]. 2nd ed. Beijing：Electronic Industry Press，2002.

[8] 吴大正，杨林耀，张永瑞，等. 信号与线性系统分析[M]. 第4版. 北京：高等教育出版社，2005.

[9] 管致中，夏恭恪，孟桥，等. 信号与线性系统分析（上册）[M]. 第4版. 北京：高等教育出版社，2004.

[10] 梁虹，梁洁，陈跃斌，等. 信号与系统分析及MATLAB实现[M]. 北京：电子工业出版社，2002.

[11] 王宝祥. 信号与系统[M]. 第3版. 北京：电子工业出版社，2010.

[12] Haykin Simon，Barry Van Veen. Signals and Systems[M]. 2nd ed. John Wiley&Sons，Inc，2003.

[13] Chen Chi-Tsong. Signals and Systems[M]. Oxford：Oxford University Press，2004.

[14] 谷源涛，等. 信号与系统——MATLAB综合实验[M]. 北京：高等教育出版社，2008.

[15] 郑君里，应启珩，杨为理. 信号与系统（上册、下册）[M]. 第3版. 北京：高等教育出版社，2011.

[16] 芮坤生. 信号分析与处理[M]. 北京：高等教育出版社，2003.

[17] 燕庆明. 信号与系统[M]. 第2版. 北京：高等教育出版社，2007.

[18] 刘百芬，张利华. 信号与系统[M]. 北京：人民邮电出版社．2012.

[19] 王景芳. 信号与系统[M]. 北京：清华大学出版社，2010.

[20] 肖志涛，等译. 信号与系统——连续与离散[M]. 北京：电子工业出版社，2005.

[21] Roberts Michael J. Signals and Systems Analysis Using Transform Methods and MATLAB[M]. McGraw-Hill，2004.

[22] Kamen Edward W，Heck Bonnie S. Fundamentals of Signal and Systems Using the Web and MATLAB[M]. Prentice Hall，2000.

[23] 陈后金，等. 信号分析与处理实验[M]. 北京：高等教育出版社，2006.